Handbook of ALGEBRA & ALGEBRAIC TOPOLOGY

Handbook of
ALGEBRA & ALGEBRAIC TOPOLOGY

Madan Khurana
R.L. Aggarwal

KNOWLEDGE BAKERS

Cataloging in Publication Data--DK
Courtesy: D.K. Agencies (P) Ltd. <docinfo@dkagencies.com>

Khurana, Madan, author.
Handbook of algebra & algebraic topology / Madan Khurana, R.L. Aggarwal.
pages cm
Includes bibliographical references and index.
ISBN 9789391502294

1. Algebra. 2. Algebraic topology. I. Aggarwal, R. L., author. II. Title. III. Title: Handbook of algebra and algebraic topology.

LCC QA155.K48 2023 | DDC 512 23

ISBN: 978-93-91502-29-4

9 789391 502294

Published by : **Knowledge Bakers**
Pune, India
Email: knowledgebakers2019@gmail.com

Digitally Printed at : **Replika Press Pvt. Ltd.**

Preface

Algebra is the branch of mathematics concerning the study of the rules of operations and relations, and the constructions and concepts arising from them, including terms, polynomials, equations and algebraic structures. Together with geometry, analysis, topology, combinatorics, and number theory, algebra is one of the main branches of pure mathematics. Elementary algebra, often part of the curriculum in secondary education, introduces the concept of variables representing numbers. Statements based on these variables are manipulated using the rules of operations that apply to numbers, such as addition. This can be done for a variety of reasons, including equation solving. Algebra is much broader than elementary algebra and studies what happens when different rules of operations are used and when operations are devised for things other than numbers. Addition and multiplication can be generalized and their precise definitions lead to structures such as groups, rings and fields, studied in the area of mathematics called abstract algebra.

Algebraic topology is a branch of mathematics which uses tools from abstract algebra to study topological spaces. The basic goal is to find algebraic invariants that classify topological spaces up to homeomorphism, though usually most classify up to homotopy equivalence. Although algebraic topology primarily uses algebra to study topological problems, using topology to solve algebraic problems is sometimes also possible. Algebraic topology, for example, allows for a convenient proof that any subgroup of a free group is again a free group. An older name for the subject was combinatorial topology, implying an emphasis on how a space X was constructed from simpler ones (the modern standard tool for such construction is the CW-complex). The basic method now applied in algebraic topology is to investigate spaces via algebraic invariants by mapping them, for example, to groups which have a great deal of manageable structure in a way that respects the relation of homeomorphism (or more general homotopy) of spaces. This allows one to recast statements about topological spaces into statements about groups, which are often easier to prove.

Two major ways in which this can be done are through fundamental groups, or more generally homotopy theory, and through homology and cohomology groups. The fundamental groups give us basic information about the structure of a topological space, but they are often nonabelian and can be difficult to work with. The fundamental group of a (finite) simplicial complex does have a finitepresentation. Homology and cohomology groups, on the other hand, are abelian and in many important cases finitely generated. Finitely generated abelian groups are completely classified and are particularly easy to work with. In general, all constructions of algebraic topology are functorial; the notions of category, functor and natural transformation originated here. Fundamental groups and homology and cohomology groups are not only *invariants* of the underlying topological space, in the sense that two topological spaces which are homeomorphic have the same associated groups, but their associated morphisms also correspond — a continuous mapping of spaces induces a group homomorphism on the associated groups, and these homomorphisms can be used to show non-existence (or, much more deeply, existence) of mappings. Several useful results follow immediately from working with finitely generated abelian groups. The free rank of the n-th homology group of a simplicial complex is equal to the n-th Betti number, so one can use the homology groups of a simplicial complex to calculate its Euler-Poincaré characteristic.

A new approach uses a functor from filtered spaces to crossed complexes defined directly and homotopically using relative homotopy groups; a higher homotopy van Kampen theorem proved for this functor enables basic results in algebraic topology, especially on the border between homology and homotopy, to be obtained without using singular homology or simplicial approximation. This approach is also called nonabelian algebraic topology, and generalises to higher dimensions ideas coming from the fundamental group.

This handbook has included more from the above like laboratory work and manipulation of concrete materials by combining the teaching of mathematics and science, with emphasis on the practical side of mathematics, that is, arithmetic computations, mechanical drawing, and graphical methods.

Dr Madan Khurana
Dr R.L. Aggarwal

Contents

Chapter 1

An Introduction

Algebra

Algebra is the branch of mathematics concerning the study of the rules of operations and relations, and the constructions and concepts arising from them, including terms, polynomials, equations and algebraic structures. Together with geometry, analysis, topology, combinatorics, and number theory, algebra is one of the main branches of pure mathematics.

Elementary algebra, often part of the curriculum in secondary education, introduces the concept of variables representing numbers. Statements based on these variables are manipulated using the rules of operations that apply to numbers, such as addition. This can be done for a variety of reasons, including equation solving. Algebra is much broader than elementary algebra and studies what happens when different rules of operations are used and when operations are devised for things other than numbers. Addition and multiplication can be generalised and their precise definitions lead to structures such as groups, rings and fields, studied in the area of mathematics called abstract algebra.

Classification

Algebra may be divided roughly into the following categories:

- Elementary algebra, in which the properties of operations on the real number system are recorded using symbols as "place holders" to denote constants and variables, and the rules governing mathematical expressions and equations involving these symbols are studied. This is usually taught at school under the title *algebra* (or *intermediate algebra* and *college algebra* in subsequent years). University-level courses in group theory may also be called *elementary algebra*.

- Abstract algebra, sometimes also called *modern algebra,* in which algebraic structures such as groups, rings and fields are axiomatically defined and investigated.
- Linear algebra, in which the specific properties of vector spaces are studied (including matrices);
- Universal algebra, in which properties common to all algebraic structures are studied.
- Algebraic number theory, in which the properties of numbers are studied through algebraic systems. Number theory inspired much of the original abstraction in algebra.
- Algebraic geometry applies abstract algebra to the problems of geometry.
- Algebraic combinatorics, in which abstract algebraic methods are used to study combinatorial questions.

In some directions of advanced study, axiomatic algebraic systems such as groups, rings, fields, and algebras over a field are investigated in the presence of a geometric structure (a metric or a topology) which is compatible with the algebraic structure. The list includes a number of areas of functional analysis:

- Normed linear spaces
- Banach spaces
- Hilbert spaces
- Banach algebras
- Normed algebras
- Topological algebras
- Topological groups

Elementary Algebra

Elementary algebra is the most basic form of algebra. It is taught to students who are presumed to have no knowledge of mathematics beyond the basic principles of arithmetic. In arithmetic, only numbers and their arithmetical operations (such as +, -, ×, ÷*) occur. In algebra, numbers are often denoted by symbols (such as* a, x, *or* y*). This is useful because:*

- It allows the general formulation of arithmetical laws (such as $a + b = b + a$ for all a and b), and thus is the first step to a systematic exploration of the properties of the real number system.

- It allows the reference to "unknown" numbers, the formulation of equations and the study of how to solve these. (For instance, "Find a number x such that $3x + 1 = 10$" or going a bit further "Find a number x such that $ax+b=c$". This step leads to the conclusion that it is not the nature of the specific numbers that allows us to solve it, but that of the operations involved.)
- It allows the formulation of functional relationships. (For instance, "If you sell x tickets, then your profit will be $3x - 10$ dollars, or $f(x) = 3x - 10$, where f is the function, and x is the number to which the function is applied.")

Polynomials

A polynomial is an expression that is constructed from one or more variables and constants, using only the operations of addition, subtraction, and multiplication (where repeated multiplication of the same variable is standardly denoted as exponentiation with a constant nonnegative integer exponent). For example, $x^2 + 2x - 3$ is a polynomial in the single variable x.

An important class of problems in algebra is factorisation of polynomials, that is, expressing a given polynomial as a product of other polynomials. The example polynomial above can be factored as $(x - 1)(x + 3)$. A related class of problems is finding algebraic expressions for the roots of a polynomial in a single variable.

Abstract Algebra

Abstract algebra extends the familiar concepts found in elementary algebra and arithmetic of numbers to more general concepts.

Sets: Rather than just considering the different types of numbers, abstract algebra deals with the more general concept of sets: a collection of all objects (called elements) selected by property, specific for the set. All collections of the familiar types of numbers are sets. Other examples of sets include the set of all two-by-two matrices, the set of all second-degree polynomials ($ax^2 + bx + c$), the set of all two dimensional vectors in the plane, and the various finite groups such as the cyclic groups which are the group of integers modulo n. Set theory is a branch of logic and not technically a branch of algebra.

Binary operations: The notion of addition (+) is abstracted to give a binary operation, * say. The notion of binary operation is meaningless without the set on which the operation is defined. For two elements a and b in a set S, a * b is another element in the set; this condition is called closure. Addition (+), subtraction (-), multiplication (×), and division (χ) can be binary operations when defined on different sets, as is addition and multiplication of matrices, vectors, and polynomials.

Identity Elements: The numbers zero and one are abstracted to give the notion of an identity element for an operation. Zero is the identity element for addition and one is the identity element for multiplication. For a general binary operator " the identity element e must satisfy a * e = a and e * a= a. This holds for addition as a + 0 = a and 0 + a = a and multiplication a × 1 = a and 1 × a = a. Not all set and operator combinations have an identity element; for example, the positive natural numbers (1, 2, 3, ...) have no identity element for addition.

Inverse Elements: The negative numbers give rise to the concept of inverse elements. For addition, the inverse of a is -a, and for multiplication the inverse is 1/a. A general inverse element a^{-1} must satisfy the property that a *a^{-} = e and a^{-1} * = e.

Associativity: Addition of integers has a property called associativity. That is, the grouping of the numbers to be added does not affect the sum. For example: (2 + 3) + 4 = 2 + (3 + 4). In general, this becomes (a * b) * c = a * (b * c). This property is shared by most binary operations, but not subtraction or division or octonion multiplication.

Commutativity: Addition and multiplication of real numbers are both commutative. That is, the order of the numbers does not affect the result. For example: 2+3=3+2. In general, this becomes a * b = b* a. This property does not hold for all binary operations. For example, matrix multiplication and quaternion multiplication are both non-commutative.

Groups

Combining the above concepts gives one of the most important structures in mathematics: a group. A group is a combination of a set S and a single binary operation *, defined in any way you choose, but with the following properties:

- An identity element e exists, such that for every member a of S, $e * a$ and $a * e$ are both identical to a.
- Every element has an inverse: for every member a of S, there exists a member a^{-1} such that $a * a^{-1}$ and $a^{-1} * a$ are both identical to the identity element.

- The operation is associative: if a, b and c are members of S, then $(a * b) * c$ is identical to $a * (b * c)$.

If a group is also commutative—that is, for any two members a and b of S, $a \cdot b$ is identical to $b \cdot a$—then the group is said to be abelian.

For example, the set of integers under the operation of addition is a group. In this group, the identity element is 0 and the inverse of any element a is its negation, $-a$. The associativity requirement is met, because for any integers a, b and c, $(a + b) + c = a + (b + c)$

The nonzero rational numbers form a group under multiplication. Here, the identity element is 1, since $1 \times a = a \times 1 = a$ for any rational number a. The inverse of a is $1/a$, since $a \times 1/a = 1$.

The integers under the multiplication operation, however, do not form a group. This is because, in general, the multiplicative inverse of an integer is not an integer. For example, 4 is an integer, but its multiplicative inverse is ¼, which is not an integer.

The theory of groups is studied in group theory. A major result in this theory is the classification of finite simple groups, mostly published between about 1955 and 1983, which is thought to classify all of the finite simple groups into roughly 30 basic types.

Semigroups, quasigroups, and monoids are structures similar to groups, but more general. They comprise a set and a closed binary operation, but do not necessarily satisfy the other conditions. Asemigroup has an *associative* binary operation, but might not have an identity element. A monoid is a semigroup which does have an identity but might not have an inverse for every element. A quasigroup satisfies a requirement that any element can be turned into any other by a unique pre- or post-operation; however the binary operation might not be associative.

All groups are monoids, and all monoids are semigroups.

Rings and Fields

Groups just have one binary operation. To fully explain the behaviour of the different types of numbers, structures with two operators need to be studied. The most important of these are rings, and fields.

A ring has two binary operations (+) and (×), with × distributive over +. Under the first operator (+) it forms an *abelian group*. Under the second operator (×) it is associative, but it does not need to have identity, or inverse, so division is not required. The additive (+)

identity element is written as 0 and the additive inverse of a is written as $-a$.

Distributivity generalises the *distributive law* for numbers, and specifies the order in which the operators should be applied, (called the precedence). For the integers $(a + b) \times c = a \times c + b \times c$ and $c \times (a + b) = c \times a + c \times b$, and $\times$ is said to be *distributive* over +.

The integers are an example of a ring. The integers have additional properties which make it an integral domain.

A field is a *ring* with the additional property that all the elements excluding 0 form an *abelian group under* $\times$. The multiplicative ($\times$) identity is written as 1 and the multiplicative inverse of a is written as a^{-1}.

The rational numbers, the real numbers and the complex numbers are all examples of fields.

Objects Called Algebras

The word algebra *is also used for various algebraic structures:*

- Algebra over a field or more generally Algebra over a ring
- Algebra over a set
- Boolean algebra
- Heyting algebra
- F-algebra and F-coalgebra in category theory
- Relational algebra
- Sigma-algebra
- T-Algebras of monads.

Topology

Topology is a major area of mathematics concerned with properties that are preserved under continuous deformations of objects, such as deformations that involve stretching, but no tearing or gluing. It emerged through the development of concepts from geometry and set theory, such as space, dimension, and transformation.

Ideas that are now classified as topological were expressed as early as 1736. Toward the end of the 19th century, a distinct discipline developed, which was referred to in Latin as the *geometria situs* ("geometry of place") or *analysis situs* (Greek-Latin for "picking apart of place"). This later acquired the modern name of topology. By the middle of the 20th century, topology had become an important area of study within mathematics.

The word *topology* is used both for the mathematical discipline and for a family of sets with certain properties that are used to define a topological space, a basic object of topology. Of particular importance are *homeomorphisms*, which can be defined as continuous functions with a continuous inverse.

Topology includes many subfields. The most basic and traditional division within topology is point-set topology, which establishes the foundational aspects of topology and investigates concepts inherent to topological spaces (basic examples include compactness and connectedness); algebraic topology, which generally tries to measure degrees of connectivity using algebraic constructs such as homotopy groups and homology; and geometric topology, which primarily studies manifolds and their embeddings (placements) in other manifolds. Some of the most active areas, such as low dimensional topology and graph theory, do not fit neatly in this division. Knot theory studies mathematical knots.

Figure 1: A three-dimensional depiction of a thickened trefoil knot, the simplest non-trivial knot

History

Topology began with the investigation of certain questions in geometry. Leonhard Euler's 1736 paper on the Seven Bridges of Kφnigsberg is regarded as one of the first academic treatises in modern topology.

The term "Topologie" was introduced in German in 1847 by Johann Benedict Listing in *Vorstudien zur Topologie*, who had used the word for ten years in correspondence before its first appearance in print. "Topology," its English form, was first used in 1883 in Listing's obituary in the journal *Nature* to distinguish "qualitative geometry from the ordinary geometry in which quantitative relations chiefly are treated".

The term topologist in the sense of a specialist in topology was used in 1905 in the magazine *Spectator*. However, none of these uses corresponds exactly to the modern definition of topology.

Modern topology depends strongly on the ideas of set theory, developed by Georg Cantor in the later part of the 19th century. Cantor, in addition to setting down the basic ideas of set theory, considered point sets in Euclidean space, as part of his study of Fourier series. Henri Poincare published *Analysis Situs* in 1895, introducing the concepts of homotopy and homology, which are now considered part of algebraic topology.

Maurice Frechet, unifying the work on function spaces of Cantor, Volterra, Arzelà, Hadamard, Ascoli, and others, introduced the metric space in 1906. A metric space is now considered a special case of a general topological space. In 1914, Felix Hausdorff coined the term "topological space" and gave the definition for what is now called a Hausdorff space. In current usage, a topological space is a slight generalisation of Hausdorff spaces, given in 1922 by Kazimierz Kuratowski.

Elementary Introduction

Topology, as a branch of mathematics, can be formally defined as "the study of qualitative properties of certain objects (called topological spaces) that are invariant under certain kind of transformations (called continuous maps), especially those properties that are invariant under a certain kind of equivalence (called homeomorphism)." To put it more simply, topology is the study of continuity and connectivity.

The term topology *is also used to refer to a structure imposed upon a set* X, *a structure that essentially 'characterises' the set* X *as a topological space by taking proper care of properties such as convergence, connectedness and continuity, upon transformation.*

Topological spaces show up naturally in almost every branch of mathematics. This has made topology one of the great unifying ideas of mathematics.

The motivating insight behind topology is that some geometric problems depend not on the exact shape of the objects involved, but rather on the way they are put together. For example, the square and the circle have many properties in common: they are both one dimensional objects (from a topological point of view) and both separate the plane into two parts, the part inside and the part outside. One of the first papers in topology was the demonstration, by Leonhard Euler, that it was impossible to find a route through the town of Kφnigsberg (now Kaliningrad) that would cross each of its seven bridges exactly once. This result did not depend on the lengths of the bridges, nor on their distance from one another, but only on connectivity properties: which bridges are connected to which islands or riverbanks. This problem, the Seven Bridges of Kφnigsberg, is now a famous problem in introductory mathematics, and led to the branch of mathematics known as graph theory.

Similarly, the hairy ball theorem of algebraic topology says that "one cannot comb the hair flat on a hairy ball without creating a cowlick." This fact is immediately convincing to most people, even though they might not recognise the more formal statement of the theorem, that there is no nonvanishing continuous tangent vector field on the sphere. As with the Bridges of Kφnigsberg, the result does not depend on the exact shape of the sphere; it applies to pear shapes and in fact any kind of smooth blob, as long as it has no holes.

To deal with these problems that do not rely on the exact shape of the objects, one must be clear about just what properties these problems do rely on. From this need arises the notion of homeomorphism. The impossibility of crossing each bridge just once applies to any arrangement of bridges homeomorphic to those in Kφnigsberg, and the hairy ball theorem applies to any space homeomorphic to a sphere.

Intuitively two spaces are homeomorphic if one can be deformed into the other without cutting or gluing. A traditional joke is that a topologist can't distinguish a coffee mug from a doughnut, since a sufficiently pliable doughnut could be reshaped to the form of a coffee cup by creating a dimple and progressively enlarging it, while shrinking the hole into a handle. A precise definition of homeomorphic, involving a continuous function with a continuous inverse, is necessarily more technical.

Homeomorphism can be considered the most basic topological equivalence. *Another is homotopy equivalence. This is harder to describe without getting technical, but the essential notion is that two objects are homotopy equivalent if they both result from "squishing" some larger object.*

Equivalence classes of the English alphabet in uppercase sans-serif font (Myriad):

Homeomorphism	*Homotopy Equivalence*
{A, R} {B} {C, G, I, J, L, M, N, S, U, V, W, Z} {D, O} {E, F, T, Y} {H, K} {P, Q} {X}	{A, R, D, O, P, Q} {B} {C, E, F, G, H, I, J, K, L, M, N, S, T, U, V, W, X, Y, Z}

An introductory exercise is to classify the uppercase letters of the English alphabet according to homeomorphism and homotopy equivalence. The result depends partially on the font used. The figures use a sans-serif font named Myriad. Notice that homotopy equivalence is a rougher relationship than homeomorphism; a homotopy equivalence class can contain several of the homeomorphism classes. The simple case of homotopy equivalence described above can be used here to show two letters are homotopy equivalent, e.g. O fits inside P and the tail of the P can be squished to the "hole" part.

Thus, the homeomorphism classes are: one hole two tails, two holes no tail, no holes, one hole no tail, no holes three tails, a bar with four tails, one hole one tail, and no holes four tails.

The homotopy classes are larger, because the tails can be squished down to a point. The homotopy classes are: one hole, two holes, and no holes.

To be sure we have classified the letters correctly, we not only need to show that two letters in the same class are equivalent, but that two letters in different classes are not equivalent. In the case of homeomorphism, this can be done by suitably selecting points and showing their removal disconnects the letters differently. For example, X and Y are not homeomorphic because removing the centre point of the X leaves four pieces; whatever point in Y corresponds to this point, its removal can leave at most three pieces. The case of homotopy equivalence is harder and requires a more elaborate argument showing an algebraic invariant, such as the fundamental group, is different on the supposedly differing classes.

Letter topology has some practical relevance in stencil typography. The font Braggadocio, for instance, has stencils that are made of one connected piece of material.

Mathematical Definition

Let X *be a set and let* τ *be a family of subsets of* X. *Then* τ *is called a* topology on X *if:*

1. Both the empty set and X are elements of τ
2. Any union of elements of τ is an element of τ

3. Any intersection of finitely many elements of τ is an element of τ

If τ is a topology on X, then the pair (X, τ) is called a *topological space*. The notation X_τ may be used to denote a set X endowed with the particular topology τ.

The members of τ are called *open sets* in X. A subset of X is said to be closed if its complement is in τ (i.e., its complement is open). A subset of X may be open, closed, both (clopen set), or neither. The empty set and X itself are always clopen.

A function or map from one topological space to another is called *continuous* if the inverse image of any open set is open. If the function maps the real numbers to the real numbers (both spaces with the Standard Topology), then this definition of continuous is equivalent to the definition of continuous in calculus. If a continuous function is one-to-one and onto, and if the inverse of the function is also continuous, then the function is called a homeomorphism and the domain of the function is said to be homeomorphic to the range. Another way of saying this is that the function has a natural extension to the topology. If two spaces are homeomorphic, they have identical topological properties, and are considered topologically the same. The cube and the sphere are homeomorphic, as are the coffee cup and the doughnut. But the circle is not homeomorphic to the doughnut.

Topology Topics

Some Theorems in General Topology:

- Every closed interval in R of finite length is compact. More is true: In R^n, a set is compact if and only if it is closed and bounded.
- Every continuous image of a compact space is compact.
- Tychonoff's theorem: the (arbitrary) product of compact spaces is compact.
- A compact subspace of a Hausdorff space is closed.
- Every continuous bijection from a compact space to a Hausdorff space is necessarily ahomeomorphism.
- Every sequence of points in a compact metric space has a convergent subsequence.
- Every interval in R is connected.
- Every compact finite-dimensional manifold can be embedded in some Euclidean space R^n.

- The continuous image of a connected space is connected.
- Every metric space is paracompact and Hausdorff, and thus normal.
- The metrisation theorems provide necessary and sufficient conditions for a topology to come from a metric.
- The Tietze extension theorem: In a normal space, every continuous real-valued function defined on a closed subspace can be extended to a continuous map defined on the whole space.
- Any open subspace of a Baire space is itself a Baire space.
- The Baire category theorem: If X is a complete metric space or a locally compact Hausdorff space, then the interior of every union of countably many nowhere dense sets is empty.
- On a paracompact Hausdorff space every open cover admits a partition of unity subordinate to the cover.
- Every path-connected, locally path-connected and semi-locally simply connected space has a universal cover.

General topology also has some surprising connections to other areas of mathematics. For example:

- In number theory, Fürstenberg's proof of the infinitude of primes.

Some Useful Notions from Algebraic Topology

- Homology and cohomology: Betti numbers, Euler characteristic, degree of a continuous mapping.
- Operations: cup product, Massey product
- Intuitively attractive applications: Brouwer fixed-point theorem, Hairy ball theorem, Borsuk–Ulam theorem, Ham sandwich theorem.
- Homotopy groups (including the fundamental group).
- Chern classes, Stiefel–Whitney classes, Pontryagin classes.

Generalisations

Occasionally, one needs to use the tools of topology but a "set of points" is not available. In pointless topology one considers instead the lattice of open sets as the basic notion of the theory, while Grothendieck topologies are certain structures defined on arbitrary categories that allow the definition of sheaves on those categories, and with that the definition of quite general cohomology theories.

Algebraic Topology

Algebraic topology is a branch of mathematics which uses tools from abstract algebra to study topological spaces. The basic goal is to find algebraic invariants that classify topological spaces up to homeomorphism, though usually most classify up to homotopy equivalence.

Although algebraic topology primarily uses algebra to study topological problems, using topology to solve algebraic problems is sometimes also possible. Algebraic topology, for example, allows for a convenient proof that any subgroup of a free group is again a free group.

The Method of Algebraic Invariants

An older name for the subject was combinatorial topology, implying an emphasis on how a space X was constructed from simpler ones (the modern standard tool for such construction is the CW-complex). The basic method now applied in algebraic topology is to investigate spaces via algebraic invariants by mapping them, for example, to groups which have a great deal of manageable structure in a way that respects the relation of homeomorphism (or more general homotopy) of spaces. This allows one to recast statements about topological spaces into statements about groups, which are often easier to prove.

Two major ways in which this can be done are through fundamental groups, or more generally homotopy theory, and through homology and cohomology groups. The fundamental groups give us basic information about the structure of a topological space, but they are often nonabelian and can be difficult to work with. The fundamental group of a (finite) simplicial complex does have a finite presentation.

Homology and cohomology groups, on the other hand, are abelian and in many important cases finitely generated. Finitely generated abelian groups are completely classified and are particularly easy to work with.

Setting in Category Theory

In general, all constructions of algebraic topology are functorial; the notions of category, functor andnatural transformation originated here. Fundamental groups and homology and cohomology groups are not only invariants *of the underlying topological space, in the sense that two topological spaces which are homeomorphic have the same associated groups, but their associated morphisms also correspond — a continuous mapping of spaces induces a group homomorphism on the associated groups, and these homomorphisms can be used to show non-existence (or, much more deeply, existence) of mappings.*

Results on Homology

Several useful results follow immediately from working with finitely generated abelian groups. The free rank of the n*-th homology group of a simplicial complex is equal to the* n*-th Betti number, so one can use the homology groups of a simplicial complex to calculate its Euler-Poincaré characteristic. As another example, the top-dimensional integral homology group of a closed manifold detects orientability: this group is isomorphic to either the integers or 0, according as the manifold is orientable or not. Thus, a great deal of topological information is encoded in the homology of a given topological space.*

Beyond simplicial homology, which is defined only for simplicial complexes, one can use the differential structure of smooth manifolds via de Rham cohomology, or Èech or sheaf cohomology to investigate the solvability of differential equations defined on the manifold in question. De Rham showed that all of these approaches were interrelated and that, for a closed, oriented manifold, the Betti numbers derived through simplicial homology were the same Betti numbers as those derived through de Rham cohomology. This was extended in the 1950s, when Eilenberg and Steenrod generalised this approach. They defined homology and cohomology as functors equipped with natural transformations subject to certain axioms (e.g., a weak equivalence of spaces passes to an isomorphism of homology groups), verified that all existing (co)homology theories satisfied these axioms, and then proved that such an axiomatisation uniquely characterised the theory.

A new approach uses a functor from filtered spaces to crossed complexes defined directly and homotopically using relative homotopy groups; a higher homotopy van Kampen theorem proved for this functor enables basic results in algebraic topology, especially on the border between homology and homotopy, to be obtained without using singular homology or simplicial approximation. This approach is also called nonabelian algebraic topology, and generalises to higher dimensions ideas coming from the fundamental group.

Applications of Algebraic Topology

Classic applications of algebraic topology include:

- The Brouwer fixed point theorem: every continuous map from the unit n-disk to itself has a fixed point.
- The n-sphere admits a nowhere-vanishing continuous unit vector field if and only if n is odd. (For$n = 2$, this is sometimes called the "hairy ball theorem".)

- The Borsuk–Ulam theorem: any continuous map from the n-sphere to Euclidean n-space identifies at least one pair of antipodal points.
- Any subgroup of a free group is free. This result is quite interesting, because the statement is purely algebraic yet the simplest proof is topological. Namely, any free group G may be realised as the fundamental group of a graph X. The main theorem on covering spaces tells us that every subgroup H of G is the fundamental group of some covering space Y of X; but every such Y is again a graph. Therefore its fundamental group H is free.

On the other hand this type of application is also handled by the use of covering morphisms of groupoids, and that technique has yielded subgroup theorems not yet proved by methods of algebraic topology.

- Topological combinatorics

Notable Algebraic Topologists

- Frank Adams
- Karol Borsuk
- Luitzen Egbertus Jan Brouwer
- William Browder
- Ronald Brown (mathematician)
- Henri Cartan
- Samuel Eilenberg
- Hans Freudenthal
- Peter Freyd
- Alexander Grothendieck
- Friedrich Hirzebruch
- Heinz Hopf
- Michael J. Hopkins
- Witold Hurewicz
- Egbert van Kampen
- Daniel Kan
- Hermann Künneth
- Solomon Lefschetz
- Jean Leray

- Saunders Mac Lane
- Mark Mahowald
- J. Peter May
- John Coleman Moore
- Jack Morava
- Goro Nishida
- Sergei Novikov
- Lev Pontryagin
- Mikhail Postnikov
- Daniel Quillen
- Jean-Pierre Serre
- Stephen Smale
- Norman Steenrod
- Dennis Sullivan
- René Thom
- Hiroshi Toda
- Leopold Vietoris
- Hassler Whitney
- J. H. C. Whitehead

Important Theorems in Algebraic Topology

- Borsuk-Ulam theorem
- Brouwer fixed point theorem
- Cellular approximation theorem
- Eilenberg–Zilber theorem
- Freudenthal suspension theorem
- Hurewicz theorem
- Künneth theorem
- Poincaré duality theorem
- Universal coefficient theorem
- Van Kampen's theorem
- Generalised van Kampen's theorems
- Higher homotopy, generalised van Kampen's theorem
- Whitehead's theorem

Chapter 2

Classification of Algebra

Elementary Algebra

Elementary algebra is a fundamental and relatively basic form of algebra taught to students who are presumed to have little or no formal knowledge of mathematics beyond arithmetic. It is typically taught in secondary school under the term algebra.

The major difference between algebra and arithmetic is the inclusion of variables. While in arithmetic only numbers and their arithmetical operations (such as +, -, ×, χ) occur, in algebra, one also uses variables such as x and y, or a and b to replace numbers.

Features of Algebra

Variables: The purpose of using variables, symbols that denote numbers, is to allow the making of generalisations in mathematics. This is useful because:

- It allows arithmetical equations (and inequalities) to be stated as laws (such as $a + b = b + a$ for all$a$ and b), and thus is the first step to the systematic study of the properties of the real number system.
- It allows reference to numbers which are not known. In the context of a problem, a variable may represent a certain value which is not yet known, but which may be found through the formulation and manipulation of equations.
- It allows the exploration of mathematical relationships between quantities (such as "if you sell $x$tickets, then your profit will be $3x$ - 10 dollars").

Expressions

In elementary algebra, an expression may contain numbers, variables and arithmetical operations. These are conventionally written with 'higher-power' terms on the left; a few examples are:

$x + 3$

$y^2 + 2x - 3$

$z^7 + a(b + x^3) + 42 / y - \pi.$

In more advanced algebra, an expression may also include elementary functions.

Operations

Properties of Operations:

Operation	***Is Written***	***commutative***	***associative***	***identity element***	***inverse operation***
Addition	$a + b$	$a + b = b + a$	$(a + b) + c = a + (b + c)$	0, which preserves numbers: $a + 0 = a$	Subtraction (-)
Multiplication	$a \times b$ or $a \cdot b$	$a \times b = b \times a$	$(a \times b) \times c = a \times (b \times c)$	1, which preserves numbers: $a \times 1 = a$	Division (/)
Exponentiation	a^b or a^b	Not commutativea^b?b^a	Not associative	1, which preserves numbers: $a^1 = a$	Logarithm (Log)

- The operation of addition...
- means repeated addition of ones: $n = 1 + 1 + ... + 1$ (n number of times);
- has an inverse operation called subtraction: $(a + b) - b = a$, which is the same as adding a negative number, $a - b = a + (-b)$;
- The operation of multiplication...
- means repeated addition: $a \times n = a + a + ... + a$ (n number of times);
- has an inverse operation called division which is defined for non-zero numbers: $(ab)/b = a$, which is the same as multiplying by a reciprocal, $a/b = a(1/b)$;

- distributes over addition: $(a + b)c = ac + bc$;
- is abbreviated by juxtaposition: $a \times b \equiv ab$;
- The operation of exponentiation...
- means repeated multiplication: $a^n = a \times a \times ... \times a$ (n number of times);
- has an inverse operation, called the logarithm: $a^{\log_a b} = b = \log_a a^b$;
- distributes over multiplication: $(ab)^c = a^c b^c$;
- can be written in terms of n-th roots: $a^{m/n} \equiv (^n\Phi a)^m$ and thus even roots of negative numbers do not exist in the real number system.
- has the property: $a^b a^c = a^{b+c}$;
- has the property: $(a^b)^c = a^{bc}$.
- in general $a^b \neq b^a$ and $(a^b)^c \neq a^{(bc)}$;

Order of Operations

In mathematics it is important that the value of an expression is always computed the same way. Therefore, it is necessary to compute the parts of an expression in a particular order, known as the order of operations. The standard order of operations is expressed in the following chart.

A common mnemonic device for remembering this order is PEMDAS. Generally in Elementary Algebra, the use of brackets (often called parentheses) and their simple applications will be taught at most of the schools in the world.

Equations

An equation is the claim that two expressions have the same value and are equal. Some equations are true for all values of the involved variables (such as $a + b = b + a$); such equations are called identities. Conditional equations are true for only some values of the involved variables: $x^2 - 1 = 4$. The values of the variables which make the equation true are the solutions of the equation and can be found through equation solving.

Properties of Equality

- The relation of equality (=) is...
- reflexive: $b = b$;
- symmetric: if $a = b$ then $b = a$;

- transitive: if $a = b$ and $b = c$ then $a = c$.
- The relation of equality (=) has the property...
- that if $a = b$ and $c = d$ then $a + c = b + d$ and $ac = bd$;
- that if $a = b$ then $a + c = b + c$;
- that if two symbols are equal, then one can be substituted for the other.

Properties of Inequality

- The relation of inequality (<) has the property...
- of transivity: if $a < b$ and $b < c$ then $a < c$;
- that if $a < b$ and $c < d$ then $a + c < b + d$;
- that if $a < b$ and $c > 0$ then $ac < bc$;
- that if $a < b$ and $c < 0$ then $bc < ac$.

Algebraic Examples

The following sections lay out examples of some of the types of alegbraic equations you might encounter.

Linear Equations in One Variable

The simplest equations to solve are linear equations that have only one variable. They contain only constant numbers and a single variable without an exponent. For example:

$$2x + 4 = 12.$$

The central technique is add, subtract, multiply, or divide both sides of the equation by the same number in order to isolate the variable on one side of the equation. Once the variable is isolated, the other side of the equation is the value of the variable. For example, by subtracting 4 from both sides in the equation above:

$$2x + 4 - 4 = 12 - 4$$

can simplify to:

$$2x = 8.$$

Dividing both sides by 2:

$$\frac{2x}{2} = \frac{8}{2}$$

simplifies to the solution:

$$x = 4.$$

The general casc,

$$ax+b = c$$

follows the same procedure to obtain the solution:

$$x = \frac{c-b}{a}$$

Quadratic Equations

Quadratic equations can be expressed in the form $ax^2 + bx + c = 0$, where a is not zero (if it were zero, then the equation would not be quadratic but linear). Because of this a quadratic equation must contain the term ax^2, which is known as the quadratic term. Hence a '$\neq 0$, and so we may divide by a and rearrange the equation into the standard form

$$x^2 + px = q$$

where $p = b/a$ and $q = -c/a$. Solving this, by a process known as completing the square, leads to the quadratic formula.

Quadratic equations can also be solved using factorisation (the reverse process of which is expansion, but for two linear terms is somctimcs denoted foiling). As an example of factoring:

$$x^2 + 3x - 10 = 0.$$

Which is the same thing as

$$(x+5)(x-2) = 0.$$

It follows from the zero-product property that either $x = 2$ or $x = -5$ are the solutions, since precisely one of the factors must be equal to zero. All quadratic equations will have two solutions in the complex number system, but need not have any in the real number system. For example,

$$x^2 + 1 = 0$$

has no real number solution since no real number squared equals -1. Sometimes a quadratic equation has a root of multiplicity 2, such as:

$$(x+1)^2 = 0.$$

For this equation, -1 is a root of multiplicity 2. This means -1 appears two times.

Exponential and Logarithmic Equations

An exponential equation is an equation of the form $a^X = b$ for $a > 0$, which has solution

$$X = \log_a b = \frac{\ln b}{\ln a}$$

when $b > 0$. Elementary algebraic techniques are used to rewrite a given equation in the above way before arriving at the solution. For example, if

$$3 \cdot 2^{x-1} + 1 = 10$$

then, by subtracting 1 from both sides of the equation, and then dividing both sides by 3 we obtain

$$2^{x-1} = 3$$

whence

$$x - 1 = \log_2 3$$

or

$$x = \log_2 3 + 1.$$

A logarithmic equation is an equation of the form log aX = b for $a >$ 0, which has solution

$$X = a^b.$$

For example, if

$$4\log_5(x-3) - 2 = 6$$

then, by adding 2 to both sides of the equation, followed by dividing both sides by 4, we get

$$\log_5(x-3) = 2$$

whence

$$x - 3 = 5^2 = 25$$

from which we obtain

$$x = 28.$$

Radical Equations

A radical equation is an equation of the form $X^{m/n} = a$, for m, n integers, which has solution

$$X = \sqrt[m]{a^n} = \left(\sqrt[m]{a}\right)^n$$

if m is odd, and solution, and

$$X = \pm\sqrt[m]{a^n} = \pm\left(\sqrt[m]{a}\right)^n$$

if m is even and $a \geq 0$.

For example, if

$$(x+5)^{2/3} = 4,$$

then

$$x+5 = \pm(\sqrt{4})^3 = \pm 8$$

$$x = -5 \pm 8$$

or

$$x = 3, -13 .$$

System of Linear Equations

In the case of a system of linear equations, like, for instance, two equations in two variables, it is often possible to find the solutions of both variables that satisfy both equations.

Elimination Method

An example of solving a system of linear equations is by using the elimination method:

$$\begin{cases} 4x + 2y = 14 \\ 2x - y = 1. \end{cases}$$

Multiplying the terms in the second equation by 2:

$$4x + 2y = 14$$

$$4x - 2y = 2.$$

Adding the two equations together to get:

$$8x = 16$$

which simplifies to

$$x = 2$$

Since the fact that $x = 2$ is known, it is then possible to deduce that $y = 3$ by either of the original two equations (by using *2* instead of x) The full solution to this problem is then

$$\begin{cases} x = 2 \\ y = 3. \end{cases}$$

Note that this is not the only way to solve this specific system; y could have been solved before x.

Second Method of Finding a Solution

Another way of solving the same system of linear equations is by substitution.

$$\begin{cases} 4x + 2y = 14 \\ 2x - y = 1. \end{cases}$$

An equivalent for *y* can be deduced by using one of the two equations. Using the second equation:

$$2x - y = 1$$

Subtracting *2x* from each side of the equation:

$$2x - 2x - y = 1 - 2x$$

$$-y = 1 - 2x$$

and multiplying by -1:

$$y = 2x - 1.$$

Using this *y* value in the first equation in the original system:

$$4x + 2(2x - 1) = 14$$

$$4x + 4x - 2 = 14$$

$$8x - 2 = 14$$

Adding *2* on each side of the equation:

$$8x - 2 + 2 = 14 + 2$$

$$8x = 16$$

which simplifies to

$$x = 2$$

Using this value in one of the equations, the same solution as in the previous method is obtained.

$$\begin{cases} x = 2 \\ y = 3. \end{cases}$$

Note that this is not the only way to solve this specific system; in this case as well, *y* could have been solved before *x*.

Other Types of Systems of Linear Equations

Unsolvable Systems: In the above example, it is possible to find a solution. However, there are also systems of equations which do not have a solution. An obvious example would be:

$$\begin{cases} x + y = 1 \\ 0x + 0y = 2 \end{cases}$$

The second equation in the system has no possible solution. Therefore, this system can't be solved. However, not all incompatible systems are recognised at first sight. As an example, the following system is studied:

$$\begin{cases} 4x + 2y = 12 \\ -2x - y = -4 \end{cases}$$

When trying to solve this (for example, by using the method of substitution above), the second equation, after adding $2x$ on both sides and multiplying by -1, results in:

$$y = -2x + 4$$

And using this value for y in the first equation:

$$4x + 2(-2x + 4) = 12$$

$$4x - 4x + 8 = 12$$

$$8 = 12$$

No variables are left, and the equality is not true. This means that the first equation can't provide a solution for the value for y obtained in the second equation.

Undetermined Systems

There are also systems which have multiple or infinite solutions, in opposition to a system with a unique solution (meaning, two unique values for x and y) For example:

$$\begin{cases} 4x + 2y = 12 \\ -2x - y = -6 \end{cases}$$

Isolating y in the second equation:

$$y = -2x + 6$$

And using this value in the first equation in the system:

$$4x + 2(-2x + 6) = 12$$

$$4x - 4x + 12 = 12$$

$$12 = 12$$

The equality is true, but it does not provide a value for x. Indeed, one can easily verify (by just filling in some values of x) that for any x there is a solution as long as $y = -2x + 6$. There are infinite solutions for this system.

Over- and Underdetermined Systems

Systems with more variables than the number of linear equations do not have a unique solution. An example of such a system is

$$\begin{cases} x + 2y = 10 \\ y - z = 2 \end{cases}$$

Such a system is called underdetermined; when trying to find a solution, one or more variables can only be expressed in relation to the other variables, but cannot be determined numerically. Incidentally, a system with a greater number of equations than variables, in which necessarily some equations are sums or multiples of others, is called over determined.

Relation between Solvability and Multiplicity

Given any system of linear equations, there is a relation between multiplicity and solvability.

If one equation is a multiple of the other (or, more generally, a sum of multiples of the other equations), then the system of linear equations is undetermined, meaning that the system has infinitely many solutions. Example:

$$\begin{cases} x + y = 2 \\ 2x + 2y = 4 \end{cases}$$

has solutions (x,y) such as (1,1), (0,2), (1.8,0.2), (4,-2), (-3000.75,3002.75), and so on.

When the multiplicity is only partial (meaning that for example, only the left hand sides of the equations are multiples, while the right hand sides are not or not by the *same* number) then the system is unsolvable. For example, in

$$\begin{cases} x + y = 2 \\ 4x + 4y = 1 \end{cases}$$

the second equation yields that $x + y = 1/4$ which is in contradiction with the first equation. Such a system is also called *inconsistent* in the language of linear algebra. When trying to solve a system of linear equations it is generally a good idea to check if one equation is a multiple of the other. If this is precisely so, the solution cannot be uniquely determined. If this is only partially so, the solution does not exist. This, however, does not mean that the equations must be multiples of each other to have a solution, as shown in the sections above; in other words: multiplicity in a system of linear equations is not a necessary condition for solvability.

Abstract Algebra

Abstract algebra is the subject area of mathematics that studies algebraic structures, such as groups, rings, fields, modules, vector spaces, and algebras. The phrase abstract algebra was coined at the turn of the 20th century to distinguish this area from what was normally referred to as algebra, the study of the rules for manipulating formulae and algebraic expressions involving unknowns and real or complex numbers, often now called elementary *algebra.* The distinction is rarely made in more recent writings.

Contemporary mathematics and mathematical physics make extensive use of abstract algebra; for example, theoretical physics draws on Lie algebras. Subject areas such as algebraic number theory, algebraic topology, and algebraic geometry apply algebraic methods to other areas of mathematics. Representation theory, roughly speaking, takes the 'abstract' out of 'abstract algebra', studying the concrete side of a given structure.

Two mathematical subject areas that study the properties of algebraic structures viewed as a whole are universal algebra and category theory. Algebraic structures, together with the associated homomorphisms, form categories. Category theory is a powerful formalism for studying and comparing different algebraic structures.

History

As in other parts of mathematics, concrete problems and examples have played important roles in the development of algebra. Through the end of the nineteenth century many, perhaps most of these problems were in some way related to the theory of algebraic equations. Major themes include:

- Solving of systems of linear equations, which led to matrices, determinants and linear algebra
- Attempts to find formulae for solutions of general polynomial equations of higher degree that resulted in discovery of groups as abstract manifestations of symmetry
- Arithmetical investigations of quadratic and higher degree

forms and diophantine equations, notably, in proving Fermat's last theorem, that directly produced the notions of a ring and ideal

Numerous textbooks in abstract algebra start with axiomatic definitions of various algebraic structures and then proceed to establish their properties, creating a false impression that somehow in algebra axioms had come first and then served as a motivation and as a basis of further study. The true order of historical development was almost exactly the opposite.

For example, the hyper complex numbers of the nineteenth century had kinematic and physical motivations but challenged comprehension. Most theories that are now recognised as parts of algebra started as collections of disparate facts from various branches of mathematics, acquired a common theme that served as a core around which various results were grouped, and finally became unified on a basis of a common set of concepts. An archetypical example of this progressive synthesis can be seen in the theory of groups.

Early Group Theory

There were several threads in the early development of group theory, in modern language loosely corresponding to *number theory*, *theory of equations*, and *geometry*.

Leonhard Euler considered algebraic operations on numbers modulo an integer, modular arithmetic, proving his generalisation of Fermat's little theorem. These investigations were taken much further by Carl Friedrich Gauss, who considered the structure of multiplicative groups of residues mod n and established many properties of cyclic and more general abelian groups that arise in this way.

In his investigations of composition of binary quadratic forms, Gauss explicitly stated the associative law for the composition of forms, but like Euler before him, he seems to have been more interested in concrete results than in general theory.

In 1870, Leopold Kronecker gave a definition of an abelian group in the context of ideal class groups of a number field, a far-reaching generalisation of Gauss's work. It appears that he did not tie it with previous work on groups, in particular, permutation groups. In 1882 considering the same question, Heinrich M. Weber realised the connection and gave a similar definition that involved the cancellation property but omitted the existence of the inverse element, which was sufficient in his context (finite groups).

Permutations were studied by Joseph Lagrange in his 1770 paper *Réflexions sur la résolution algébrique des équations* (*Thoughts on Solving Algebraic Equations*) devoted to solutions of algebraic equations, in which he introduced Lagrange resolvents. Lagrange's goal was to understand why equations of third and fourth degree admit formulae for solutions, and he identified as key objects permutations of the roots.

An important novel step taken by Lagrange in this paper was the abstract view of the roots, i.e. as symbols and not as numbers. However, he did not consider composition of permutations. Serendipitously, the first edition of Edward Waring's *Meditationes Algebraicae* (*Meditations on the Algebraic*) appeared in the same year, with an expanded version published in 1782. Waring proved the main theorem on symmetric functions, and specially considered the relation between the roots of a quartic equation and its resolvent cubic. *Mémoire sur la résolution des équations* (*Memoire on the Solving of Equations*) of Alexandre Vandermonde (1771) developed the theory of symmetric functions from a slightly different angle, but like Lagrange, with the goal of understanding solvability of algebraic equations.

Kronecker claimed in 1888 that the study of modern algebra began with this first paper of Vandermonde. Cauchy states quite clearly that Vandermonde had priority over Lagrange for this remarkable idea, which eventually led to the study of group theory.

Paolo Ruffini was the first person to develop the theory of permutation groups, and like his predecessors, also in the context of solving algebraic equations. His goal was to establish the impossibility of an algebraic solution to a general algebraic equation of degree greater than four. En route to this goal he introduced the notion of the order of an element of a group, conjugacy, the cycle decomposition of elements of permutation groups and the notions of primitive and imprimitive and proved some important theorems relating these concepts, such as

> *if G is a subgroup of S_5 whose order is divisible by 5 then G contains an element of order 5.*

Note, however, that he got by without formalising the concept of a group, or even of a permutation group. The next step was taken by Évariste Galois in 1832, although his work remained unpublished until 1846, when he considered for the first time what is now called the *closure property* of a group of permutations, which he expressed as:

... if in such a group one has the substitutions S and T then one has the substitution ST.

The theory of permutation groups received further far-reaching development in the hands of Augustin Cauchy and Camille Jordan, both through introduction of new concepts and, primarily, a great wealth of results about special classes of permutation groups and even some general theorems. Among other things, Jordan defined a notion of isomorphism, still in the context of permutation groups and, incidentally, it was he who put the term *group* in wide use.

The abstract notion of a group appeared for the first time in Arthur Cayley's papers in 1854. Cayley realised that a group need not be a permutation group (or even *finite*), and may instead consist of matrices, whose algebraic properties, such as multiplication and inverses, he systematically investigated in succeeding years. Much later Cayley would revisit the question whether abstract groups were more general than permutation groups, and establish that, in fact, any group is isomorphic to a group of permutations.

Modern Algebra

The end of the 19th and the beginning of the 20th century saw a tremendous shift in the methodology of mathematics. Abstract algebra emerged around the start of the 20th century, under the name modern *algebra*. Its study was part of the drive for more intellectual rigor in mathematics. Initially, the assumptions in classical algebra, on which the whole of mathematics (and major parts of the natural sciences) depend, took the form of axiomatic systems.

No longer satisfied with establishing properties of concrete objects, mathematicians started to turn their attention to general theory. Formal definitions of certain algebraic structures began to emerge in the 19th century.

For example, results about various groups of permutations came to be seen as instances of general theorems that concern a general notion of an *abstract group*. Questions of structure and classification of various mathematical objects came to forefront. These processes were occurring throughout all of mathematics, but became especially pronounced in algebra. Formal definition through primitive operations and axioms were proposed for many basic algebraic structures, such as groups, rings, and fields. Hence such things as group theory and ring theory took their places in pure mathematics. The algebraic investigations of general fields by Ernst Steinitz and of commutative and then general rings by David Hilbert, Emil Artin and Emmy Noether, building up on the work of Ernst Kummer, Leopold Kronecker and Richard Dedekind, who had considered ideals in commutative rings, and of Georg Frobenius and Issai Schur, concerning representation theory of groups, came to define abstract algebra. These developments of the last quarter of the 19th century and the first quarter of 20th century were systematically exposed in Bartel van der Waerden's *Moderne algebra*, the two-volume monograph published in 1930–1931 that forever changed for the mathematical world the meaning of the word *algebra* from *the theory of equations* to the *theory of algebraic structures*.

Examples

Abstract algebra facilitates the study of properties and patterns that seemingly disparate mathematical concepts have in common. For example, consider the distinct operations of function composition, f (g(x)), and of matrix multiplication, AB.

These two operations have, in fact, the same structure. Think about multiplying two square matrices, AB, by a one column vector, x. This defines a function equivalent to composing Ay with Bx: Ay = A(Bx) = (AB)x. Functions under composition and matrices under multiplication are examples of monoids. A set S and a binary operation over S, denoted by concatenation, form a monoid if the operation associates, (ab)c = a(bc), and if there exists an ≥S, such that ae = ea = a.

Examples of algebraic structures with a single binary operation are:

- Magmas
- Quasigroups
- Monoids
- Semigroups
- Groups.

More complicated examples include:

- Rings
- Fields
- Modules
- Vector spaces
- Algebras over fields

- Associative algebras
- Lie algebras
- Lattices
- Boolean algebras.

Linear Algebra

Linear algebra is a branch of mathematics that studies vector spaces, also called linear spaces, along with linear functions that input one vector and output another. Such functions are called linear maps (or linear transformations or linear operators) and can be represented by matrices if a basis is given. Thus matrix theory is often considered as a part of linear algebra. Linear algebra is commonly restricted to the case of finite dimensional vector spaces, while the peculiarities of the infinite dimensional case are traditionally covered in linear functional analysis. Linear algebra is central to modern mathematics and its applications. An elementary application of linear algebra is to find the solution of a system of linear equations in several unknowns. More advanced applications are ubiquitous in areas as diverse as abstract algebra and functional analysis.

Linear algebra has a concrete representation in analytic geometry and is generalised in operator theory and in module theory. It has extensive applications in engineering, physics, natural sciences, computer science, and the social sciences (particularly in economics). Nonlinear mathematical models can often be approximated by linear ones.

History

The subject first took its modern form in the first half of the twentieth century. At this time, many ideas and methods of previous centuries were generalised as abstract algebra. Matrices and tensors were introduced in the latter part of the 19th century. The use of these objects in quantum mechanics, special relativity, and statistics did much to spread the subject of linear algebra beyond pure mathematics.

The origin of many of these ideas is discussed in the articles on determinants and Gaussian elimination.

Main Structures

The main structures of linear algebra are vector spaces and linear maps between them. A vector space is a set whose elements can be added together and multiplied by the scalars, or numbers. In many physical applications, the scalars are real numbers, R. More generally, the scalars may form any field F—thus one can consider vector spaces over the field Q of rational numbers, the field C of complex numbers, or a finite field F_q.

In a vector space, the operations of addition and scalar multiplication must behave similarly to the usual addition and multiplication of numbers: addition is commutative and associative, multiplication distributes over addition, and so on. More precisely, the two operations must satisfy a list of axioms chosen to emulate the properties of addition and scalar multiplication of Euclidean vectors in the coordinate n-space R^n. One of the axioms stipulates the existence of a zero vector, which behaves analogously to the number zero with respect to addition.

Elements of a general vector space V may be objects of any nature, for example, functions or polynomials, but when viewed as elements of V, they are frequently called vectors.

Given two vector spaces V and W over a field F, a linear transformation (or "linear map") is a map

$$T : V \rightarrow W$$

that is compatible with addition and scalar multiplication:

$$T(u + v) = T(u) + T(v), \quad T(rv) = rT(v)$$

for any vectors *u,v* - *V* and a scalar *r* - *F*.

Other fundamental notions in linear algebra include: linear combination, span, linear independence of vectors, a basis of a vector space, and the dimension of a vector space.

Given a vector space *V* over a field F, an expression of the form

$$r_1 v_1 + r_2 v_2 + \cdots + r_k v_k,$$

where v_1, v_2, ..., v_k are vectors and r_1, r_2, ..., r_k are scalars, is called the linear combination of the vectors v_1, v_2, ..., v_k with coefficients r_1, r_2, ..., r_k. The set of all linear combinations of vectors v_1, v_2, ..., v_k is called their span. A linear combination of any system of vectors with all zero coefficients is zero vector of *V*. If this is the only way to express zero vector as a linear combination of v_1, v_2, ..., *vk then* these vectors are linearly independent. A linearly independent set of vectors that spans a vector space *V* is a basis of *V*. If a vector space admits a finite basis then any two bases have the same number of elements (called the dimension of *V*) and *V* is a finite-dimensional vector space. This theory can be extended to infinite-dimensional spaces.

There is an important distinction between the coordinate n-space R^n and a general finite-dimensional vector space V. While R^n has a standard basis $\{e_1, e_2, ..., e_n\}$, a vector space V typically does not come equipped with a basis and many different bases exist (although they all consist of the same number of elements equal to the dimension of V). Having a particular basis $\{v_1, v_2, ..., v_n\}$ of V allows one to construct a coordinate system in V: the vector with coordinates $(r_1, r_2, ..., r_n)$ is the linear combination

$$r_1v_1 + r_2v_2 + \cdots + r_nv_n.$$

The condition that $v_1, v_2, ..., v_n$ span V guarantees that each vector v can be assigned coordinates, whereas the linear independence of $v_1, v_2, ..., v_n$ further assures that these coordinates are determined in a unique way (i.e. there is only one linear combination of the basis vectors that is equal to v). In this way, once a basis of a vector space V over F has been chosen, V may be identified with the coordinate n-space F^n.

Under this identification, addition and scalar multiplication of vectors in V correspond to addition and scalar multiplication of their coordinate vectors in F^n. Furthermore, if V and W are an n-dimensional and m-dimensional vector space over F, and a basis of V and a basis of W have been fixed, then any linear transformation $T: V \to W$ may be encoded by an $m \times n$ matrix A with entries in the field F, called the matrix of T with respect to these bases.

Therefore, by and large, the study of linear transformations, which were defined axiomatically, may be replaced by the study of matrices, which are concrete objects. This is a major technique in linear algebra.

Some Main Useful Theorems

- (AC) Every vector space has a basis.
- (AC) Any two bases of the same vector space have the same cardinality. Equivalently, the dimension of a vector space is well-defined.
- A matrix is invertible, or non-singular, if and only if the linear map represented by the matrix is anisomorphism.
- Any vector space over a field F of dimension n is isomorphic to F^n as a vector space over F.
- Corollary: Any two vector spaces over F of the same finite dimension are isomorphic to each other.

Generalisations and Related Topics

Since linear algebra is a successful theory, its methods have been developed in other parts of mathematics. In module theory, one replaces the field of scalars by a ring. In multilinear algebra, one considers multivariable linear transformations, that is, mappings that are linear in each of a number of different variables. This line of inquiry naturally leads to the idea of the tensor product. Functional analysis mixes the methods of linear algebra with those of mathematical analysis.

Universal Algebra

Universal algebra (sometimes called general algebra) is the field of mathematics that studies algebraic structures themselves, not examples ("models") of algebraic structures. For instance, rather than take particular groups as the object of study, in universal algebra one takes "the theory of groups" as an object of study.

Basic Idea

From the point of view of universal algebra, an algebra (or algebraic structure) is a set A *together with a collection of operations on* A. *An* n-*ary operation on* A *is a function that takes* n *elements of* A*and returns a single element of* A. *Thus, a 0-ary operation (or* nullary operation*) can be represented simply as an element of* A, *or a* constant, *often denoted by a letter like* a. *A 1-ary operation (or* unary operation*) is simply a function from* A *to* A, *often denoted by a symbol placed in front of its argument, like* ~x. *A 2-ary operation (or* binary operation*) is often denoted by a symbol placed between its arguments, like* x * y. *Operations of higher or unspecified arity are usually denoted by function symbols, with the arguments placed in parentheses and separated by commas, like* $f(x,y,z)$ *or*$f(x_1,\ldots,x_n)$. *Some researchers allow infinitary operations, such as* $\bigwedge_{\alpha\in J} x_\alpha$ *where* J *is an infinite index set, thus leading into the algebraic theory of complete lattices. One way of talking about an algebra, then, is by referring to it as an algebra of a certain type* Ω, *where* Ω *is an ordered sequence of natural numbers representing the arity of the operations of the algebra.*

Equations

After the operations have been specified, the nature of the algebra can be further limited by axioms, which in universal algebra often take the form of identities, or equational laws. An example is the associative axiom for a binary operation, which is given by the equation $x * (y * z) = (x * y) * z$. The axiom is intended to hold for all elements x, y, and z of the set A.

Varieties

An algebraic structure which can be defined by identities is called a variety, and these are sufficiently important that some authors consider varieties the only object of study in universal algebra, while others consider them an object.

Restricting one's study to varieties rules out:

- Predicate logic, notably quantification, including existential quantification ($\exists$) and universal quantification ($\forall$)
- Relations, including inequalities, both $a \neq b$ and order relations

In this narrower definition, universal algebra can be seen as a special branch of model theory, in which we are typically dealing with structures having operations only (i.e. the type can have symbols for functions but not for relations other than equality), and in which the language used to talk about these structures uses equations only.

Not all algebraic structures in a wider sense fall into this scope. For example ordered groups are not studied in mainstream universal algebra because they involve an ordering relation.

A more fundamental restriction is that universal algebra cannot study the class of fields, because there is no type in which all field laws can be written as equations (inverses of elements are defined for all *non-zero* elements in a field, so inversion cannot simply be added to the type).

One advantage of this restriction is that the structures studied in universal algebra can be defined in any category which has *finite products*. For example, a topological group is just a group in the category of topological spaces.

Examples

Most of the usual algebraic systems of mathematics are examples of varieties, but not always in an obvious way – the usual definitions often involve quantification or inequalities.

Groups

Let's consider the definition of a group. Normally a group is defined in terms of a single binary operation *, subject to these axioms:

- Associativity (as in the previous section): $x * (y * z) = (x * y) * z$.
- Identity element: There exists an element e such that for each element x, $e * x = x = x * e$.
- Inverse element: It can easily be seen that the identity element is unique. If we denote this unique identity element by e then for each x, there exists an element i such that $x * i = e = i * x$.

Now, this definition of a group is problematic from the point of view of universal algebra. The reason is that the axioms of the identity element and inversion are not stated purely in terms of equational laws but also have clauses involving the phrase "there exists ... such that ...". This is inconvenient; the list of group properties can be simplified to universally quantified equations if we add a nullary operation e *and* a unary operation ~ in addition to the binary operation *, then list the axioms for these three operations as follows:

- Associativity: $x * (y * z) = (x * y) * z$.
- Identity element: $e * x = x = x * e$.
- Inverse element: $x * (\sim x) = e = (\sim x) * x$.

(Of course, we usually write "x^{-1}" instead of "$\sim x$", which shows that the notation for operations of low arity is not *always* as given in the second paragraph.)

What has changed is that in the usual definition there are:

- a single binary operation (signature (2))
- 1 equational law (associativity)
- 2 quantified laws (identity and inverse)

...while in the universal algebra definition there are

- 3 operations: one binary, one unary, and one nullary (signature (2,1,0))
- 3 equational laws (associativity, identity, and inverse)
- no quantified laws

It is important to check that this really does capture the definition of a group. The reason that it might not is that specifying one of these universal groups might give more information than specifying one of the usual kind of group. After all, nothing in the usual definition said that the identity element e was unique; if there is another identity element e', then it is ambiguous which one should be the value of the nullary operator e. However, this is not a problem because identity elements can be proved to be always unique. The same thing is true of inverse elements. So the universal algebraist's definition of a group really is equivalent to the usual definition.

Basic Constructions

We assume that the type, Ω, has been fixed. Then there are three basic constructions in universal algebra: homomorphic image, subalgebra, and product.

A homomorphism between two algebras A and B is a function $h: A \to B$ from the set A to the set B such that, for every operation *f (of arity, say, n), h* (fA(*x*1,...,*xn*)) = $f_B(h(x_1),...,h(x_n))$. (Here, subscripts are placed on f to indicate whether it is the version of f in A or B. In theory, you could tell this from the context, so these subscripts are usually left off.) For example, if e is a constant (nullary operation), then $h(e_A) = e_B$. If ~ is a unary operation, then $h(\sim x) = \sim h(x)$. If * is a binary operation, then h(x * y) =h(x) * h(y). And so on. A few of the things that can be done with homomorphisms, as well as definitions of certain special kinds of homomorphisms, are listed under the entry Homomorphism. In particular, we can take the homomorphic image of an algebra, h(A).

A subalgebra of A *is a subset of* A *that is closed under all the operations of* A. *A product of some set of algebraic structures is the cartesian product of the sets with the operations defined coordinate wise.*

Some Basic Theorems

- The Isomorphism theorems, which encompass the isomorphism theorems of groups, rings, modules, etc.
- Birkhoff's HSP Theorem, which states that a class of algebras is a variety if and only if it is closed under homomorphic images, subalgebras, and arbitrary direct products.

Motivations and Applications

In addition to its unifying approach, universal algebra also gives deep theorems and important examples and counter examples. It provides a useful framework for those who intend to start the study of new classes of algebras. It can enable the use of methods invented for some particular classes of algebras to other classes of algebras, by recasting the methods in terms of universal algebra (if possible), and then interpreting these as applied to other classes. It has also provided conceptual clarification; as J.D.H. Smith puts it, "What looks messy and complicated in a particular framework may turn out to be simple and obvious in the proper general one."

In particular, universal algebra can be applied to the study of monoids, rings, and lattices. Before universal algebra came along, many theorems (most notably the isomorphism theorems) were proved separately in all of these fields, but with universal algebra, they can be proven once and for all for every kind of algebraic system.

The 1956 paper by Higgins referenced below has been well followed up for its framwork for a range of particular algebraic systems, while his 1963 paper is notable for its discussion of algebras with operations which are only partially defined, typical examples for this being categories and groupoids. This leads on to the subject of higher dimensional algebra which can be defined as the study of algebraic theories with partial operations whose domains are defined under geometric conditions. Notable examples of these are various forms of higher dimensional categories and groupoids.

Category Theory and Operads

A more generalised programme along these lines is carried out by category theory. Given a list of operations and axioms in universal algebra, the corresponding algebras and homomorphisms are the objects and morphisms of a category. Category theory applies to many situations where universal algebra does not, extending the reach of the theorems. Conversely, many theorems that hold in universal algebra do not generalise all the way to category theory. Thus both fields of study are useful.

A more recent development in category theory that generalises operations is operad theory – an operad is a set of operations, similar to a universal algebra.

History

In Alfred North Whitehead's book A Treatise on Universal Algebra, *published in 1898, the term universal* algebra *had essentially the same meaning that it has today. Whitehead credits William Rowan Hamilton and Augustus De Morgan as originators of the subject matter, and James Joseph Sylvester with coining the term itself.*

At the time structures such as Lie algebras and hyperbolic quaternions drew attention to the need to expand algebraic structures beyond the associatively multiplicative class. In a review Alexander Macfarlane wrote: "The main idea of the work is not unification of the several methods, nor generalisation of ordinary algebra so as to include them, but rather the comparative study of their several structures." At the time George Boole's algebra of logic made a strong counterpoint to ordinary number algebra, so the term "universal" served to calm strained sensibilities.

Whitehead's early work sought to unify quaternions (due to Hamilton), Grassmann's Ausdehnungslehre, and Boole's algebra of logic. Whitehead wrote in his book:

> *"Such algebras have an intrinsic value for separate detailed study; also they are worthy of comparative study, for the sake of the light thereby thrown on the general theory of symbolic reasoning, and on algebraic symbolism in particular. The comparative study necessarily presupposes some previous separate study, comparison being impossible without knowledge."*

Whitehead, however, had no results of a general nature. Work on the subject was minimal until the early 1930s, when Garrett Birkhoff and Øystein Ore began publishing on universal algebras. Developments in metamathematics and category theory in the 1940s and 1950s furthered the field, particularly the work of Abraham Robinson, Alfred Tarski, Andrzej Mostowski, and their students (Brainerd 1967).

In the period between 1935 and 1950, most papers were written along the lines suggested by Birkhoff's papers, dealing with free algebras, congruence and subalgebra lattices, and homomorphism theorems. Although the development of mathematical logic had made applications to algebra possible, they came about slowly; results published by Anatoly Maltsev in the 1940s went unnoticed because of the war. Tarski's lecture at the 1950 International Congress of Mathematicians in Cambridge ushered in a new period in which model-theoretic aspects were developed, mainly by Tarski himself, as well as C.C. Chang, Leon Henkin, Bjarni Jónsson, Roger Lyndon, and others.

In the late 1950s, Edward Marczewski emphasized the importance of free algebras, leading to the publication of more than 50 papers on the algebraic theory of free algebras by Marczewski himself, together with Jan Mycielski, W³adys³aw Narkiewicz, Witold Nitka, J. P³onka, S. Œwierczkowski, K. Urbanik, and others.

Graph Algebra

In mathematics, especially in the fields of universal algebra and graph theory, a graph algebra is a way of giving a directed graph an algebraic structure. It was introduced in (McNulty & Shannon

1983), and has seen many uses in the field of universal algebra since then.

Definition

Let D = (V,E) *be a directed graph, and let 0 be an element not in* V. *The graph algebra associated with* D *is the set* $V \cup \{0\}$ *equipped with multiplication defined by the rules* $xy = x$ *if* $x,y \in V,(x,y) \in E$, *and* $xy = 0$ *if* $x,y \in V \cup \{0\},(x,y) \notin E$.

Applications

This notion has made it possible to use the methods of graph theory in universal algebra and several other directions of discrete mathematics and computer science. Graph algebras have been used, for example, in constructions concerning dualities (Davey et al. 2000), equational theories (Pφschel 1989), flatness (Deliæ 2001), groupoid rings (Lee 1991), topologies (Lee 1988), varieties (Oates-Williams 1984), finite state automata (Kelarev, Miller & Sokratova 2005), finite state machines (Kelarev & Sokratova 2003), tree languages and tree automata (Kelarev & Sokratova 2001) etc.

Group Algebra

In mathematics, the group algebra is any of various constructions to assign to a locally compact group an operator algebra (or more generally a Banach algebra), such that representations of the algebra are related to representations of the group. As such, they are similar to the group ring associated to a discrete group.

Group Algebras of Topological Groups: $C_c(G)$

For the purposes of functional analysis, and in particular of harmonic analysis, one wishes to carry over the group ring construction to topological groups G. In case G is a locally compact Hausdorff group, G carries an essentially unique left-invariant countably additive Borel measure μ called Haar measure. Using the Haar measure, one can define a convolution operation on the space $C_c(G)$ of complex-valued continuous functions on G with compact support; $C_c(G)$ can then be given any of various norms and the completion will be a group algebra.

To define the convolution operation, let f and g be two functions in $C_c(G)$. For t in G, define

$$[f * g](t) = \int_G f(s)g(s^{-1}t)d\mu(s).$$

The fact that $f * g$ is continuous is immediate from the dominated convergence theorem. Also

$$\text{Support}(f * g) \subseteq \text{Support}(f) \cdot \text{Support}(g)$$

$C_c(G)$ also has a natural involution defined by:

$$f^*(s) = \overline{f(s^{-1})}\Delta(s^{-1})$$

where Δ is the modular function on G. With this involution, it is a *-algebra.

Theorem. If $C_c(G)$ is given the norm

$||f||_1 := \int_G |f(s)| \, d\mu(s)$, it becomes is an involutive normed algebra with an approximate identity.

The approximate identity can be indexed on a neighbourhood basis of the identity consisting of compact sets. Indeed if V is a compact neighbourhood of the identity, let f_V be a non-negative continuous function supported in V such that

$$\int_V f_V(g) \, d\mu(g) = 1.$$

Then $\{f_V\}_V$ is an approximate identity. A group algebra can only have an identity, as opposed to just approximate identity, if and only if the topology on the group is the discrete topology.

Note that for discrete groups, $C_c(G)$ is the same thing as the complex group ring $\mathbb{C}G$.

The importance of the group algebra is that it captures the unitary representation theory of G as shown in the following

Theorem. Let G be a locally compact group. If U is a strongly continuous unitary representation of Gon a Hilbert space H, then

$$\pi_U(f) = \int_G f(g) U(g) \, d\mu(g)$$

is a non-degenerate bounded *-representation of the normed algebra $C_c(G)$. The map

$$U \mapsto \pi_U$$

is a bijection between the set of strongly continuous unitary representations of G and non-degenerate bounded *-representations of $C_c(G)$. This bijection respects unitary equivalence and strong containment. In particular, π_U is irreducible if and only if U is irreducible.

Non-degeneracy of a representation π of $C_c(G)$ on a Hilbert space H_φ means that

$$\{\pi(f)\xi : f \in C_c(G), \xi \in H_\pi\}$$

is dense in H_π.

The Convolution Algebra $L^1(G)$

It is a standard theorem of measure theory that the completion of $C_c(G)$ *in the* $L^1(G)$ *norm is isomorphic to the space* $L^1(G)$ *of equivalence classes of functions which are integrable with respect to the Haar measure, where, as usual, two functions are regarded as equivalent if and only if they differ on a set of Haar measure zero.*

Theorem. $L^1(G)$ *is a Banach *-algebra with the convolution product and involution defined above and with the* L^1 *norm.* $L^1(G)$ *also has a bounded approximate identity.*

The group C*-Algebra C*(G)

Let C[G] be the group ring of a discrete group G.

For a locally compact group G, *the group C*-algebra* C*(G) *of* G *is defined to be the C*-enveloping algebra of* $L^1(G)$, *i.e. the completion of* $C_c(G)$ *with respect to the largest C*-norm:*

$$||f||_{C^*} := \sup_\pi ||\pi(f)||,$$

where π ranges over all non-degenerate *-representations of $C_c(G)$ on Hilbert spaces. When G is discrete, it follows from the triangle inequality that, for any such π, $\pi(f) \leq ||f||_1$. So the norm is well-defined.

It follows from the definition that $C^*(G)$ has the following universal property: any *-homomorphism from C[G] to some B($\mathcal{H}$) (the C^*-algebra of bounded operators on some Hilbert space $\mathcal{H}$) factors through the inclusion map C[G] $\rightarrow$ *C*max(G)*.

The Reduced Group C*-Algebra $C^*_r(G)$

The reduced group C*-algebra $C^*_r(G)$ is the completion of $C_c(G)$ with respect to the norm

$$||f||_{C_r^*} := \sup\{||f * g||_2 : ||g||_2 = 1\},$$

where

$$||f||_2 = \sqrt{\int_G |f|^2 \, d\mu}$$

is the L^2 norm. Since the completion of $C_c(G)$ with regard to the L^2 norm is a Hilbert space, the C^*_rnorm is the norm of the bounded operator "convolution by f" acting on $L^2(G)$ and thus a C*-norm.

Equivalently, $C^*_r(G)$ is the C*-algebra generated by the image of the left regular representation on $l^2(G)$.

In general, $C^*_r(G)$ is a quotient of $C^*(G)$. The reduced group C*-algebra is isomorphic to the non-reduced group C*-algebra defined above if and only if G is amenable.

von Neumann Algebras Associated to Groups

The group von Neumann algebra W*(G) of G is the enveloping von Neumann algebra of C*(G).

For a discrete group G, we can consider the Hilbert space $l^2(G)$ for which G is an orthonourmal basis. Since G operates on $l^2(G)$ by permuting the basis vectors, we can identify the complex group ring C G with a subalgebra of the algebra of bounded operators on $l^2(G)$. The weak closure of this subalgebra, NG, is a von Neumann algebra.

The centre of NG can be described in terms of those elements of G whose conjugacy class is finite. In particular, if the identity element of G is the only group element with that property (that is, G has the infinite conjugacy class property), the centre of NG consists only of complex multiples of the identity.

NG is isomorphic to the hyperfinite type II_1 factor if and only if G is countable, amenable, and has the infinite conjugacy class property.

Incidence Algebra

In order theory, a field of mathematics, an incidence algebra is an associative algebra, defined for any locally finite partially ordered set and commutative ring with unity.

Definition

A locally finite poset is one for which every closed interval

$$[a,\ b] = \{x : a \leq x \leq b\}$$

within it is finite.

The members of the incidence algebra are the functions f assigning to each nonempty interval $[a, b]$ a scalar $f(a, b)$, which is taken from the *ring of scalars*, a commutative ring with unity. On this underlying set one defines addition and scalar multiplication point wise, and "multiplication" in the incidence algebra is a convolution defined by

$$(f * g)(a,b) = \sum_{a \le x \le b} f(a,x)g(x,b).$$

An incidence algebra is finite-dimensional if and only if the underlying poset is finite.

Related Concepts

An incidence algebra is analogous to a group algebra; indeed, both the group algebra and the incidence algebra are special cases of a categorical algebra, defined analogously; groups and posets being special kinds of categories.

Special Elements

The multiplicative identity element of the incidence algebra is the delta function, defined by

$$\delta(a,b) = \begin{cases} 1, & \text{if } a = b \\ 0, & \text{if } a < b \end{cases}$$

The zeta function of an incidence algebra is the constant function $\zeta(a, b) = 1$ for every interval $[a, b]$. Multiplying by ζ is analogous to integration. One can show that ζ is invertible in the incidence algebra (with respect to the convolution defined above). (Generally, a member h of the incidence algebra is invertible if and only if $h(x, x)$ is invertible for every x.) The multiplicative inverse of the zeta function is the Mφbius function $\mu(a, b)$; every value of $\mu(a, b)$ is an integral multiple of 1 in the base ring.

The Mφbius function can also be defined directly, by the following relation:

$$\mu(x,y) = \begin{cases} 1, & \text{if } \quad x = y \\ -\sum_{x \le z < y} \mu(x,z), & \text{for } \quad x < y \\ 0, & \text{otherwise} \end{cases}$$

Multiplying by μ is analogous to differentiation, and is called Mφbius inversion.

Examples

- Positive integers ordered by divisibility

The Mφbius function is $\mu(a, b) = \mu(b/a)$, where the second "μ" is the classical Mφbius function introduced into number theory in the 19th century.

- Finite subsets of some set E, ordered by inclusion

The Mφbius function is

$$\mu(S,T) = (-1)^{|T \setminus S|}$$

whenever S and T are finite subsets of E with $S \subseteq T$, and Mφbius inversion is called the principle of inclusion-exclusion.

Geometrically, this is a hypercube: $2^E = \{0,1\}^E$.

- Natural numbers with their usual order

The Mφbius function is

$$\mu(x,y) = \begin{cases} 1 & \text{if } y-x=0, \\ -1 & \text{if } y-x=1, \\ 0 & \text{if } y-x>1. \end{cases}$$

and Mφbius inversion is called the (backwards) difference operator.

Geometrically, this corresponds to the discrete number line.

Recall that convolution of sequences corresponds to multiplication of formal power series.

The Mφbius function corresponds to the sequence (1, -1, 0, 0, 0, ...) of coefficients of the formal power series 1 - z, and the zeta function in this case corresponds to the sequence of coefficients (1, 1, 1, 1, ...) of the formal power series $(1-z)^{-1} = 1+z+z^2+z^3+\cdots$, which is inverse. The delta function in this incidence algebra similarly corresponds to the formal power series 1.

- Subgroups of a finite p-group G, ordered by inclusion

The Mφbius function is

$\mu_G(H_1,H_2) = (-1)^k p^{\binom{k}{2}}$ if H_1 is a normal subgroup of H_2 and $H_2 / H_1 \cong (Z/pZ)^k$

and it is 0 otherwise. This is a theorem of Weisner (1935).

Finite sub-multisets of some multiset E, ordered by inclusion : The above three examples can be unified and generalised by considering a multiset E, and finite sub-multisets S and T of E. The Mφbius function is

$$\mu(S,T) = \begin{cases} 0 & \text{if } T \setminus S \text{ is a proper multiset (has repeated elements)} \\ (-1)^{|T \setminus S|} & \text{if } T \setminus S \text{ is a set (has no repeated elements)} \end{cases}$$

This generalises the positive integers ordered by divisibility by a positive integer corresponding to its multiset of prime divisors with multiplicity, e.g., 12 corresponds to the multiset {2,2,3}.

This generalises the natural numbers with their usual order by a natural number corresponding to a multiset of one underlying element and cardinality equal to that number, e.g., 3 corresponds to the multiset {1,1,1}.

Partitions of a set : Partially order the set of all partitions of a finite set by saying $\sigma \le \tau$ if σ is a finer partition than τ. Then the Mφbius function is

$$\mu(\sigma,\tau) = (-1)^{n-r}(2!)^{r_3}(3!)^{r_4}\cdots((n-1)!)^{r_n}$$

where n is the number of blocks in the finer partition σ, r is the number of blocks in the coarser partition τ, and r_i is the number of blocks of τ that contain exactly i blocks of σ.

Euler Characteristic

A poset is *bounded* if it has smallest and largest elements, which we call 0 and 1 respectively (not to be confused with the 0 and 1 of the ring of scalars). The Euler characteristic of a bounded finite poset is $\mu(0,1)$; it is always an integer. This concept is related to the classical Euler characteristic.

Reduced Incidence Algebras

Any member of an incidence algebra that assigns the same value to any two intervals that are isomorphic to each other as posets is a member of the reduced incidence algebra. This is a subalgebra of the incidence algebra, and it clearly contains the incidence algebra's identity element and zeta function. Any element of the reduced incidence algebra that is invertible in the larger incidence algebra has its inverse in the reduced incidence algebra. As a consequence, the Mφbius function is always a member of the reduced incidence algebra. Reduced incidence algebras shed light on the theory of generating functions, as alluded to in the case of the natural numbers above.

Quiver

In mathematics, a quiver is a directed graph where loops and multiple arrows between two verticesare allowed, i.e. a multidigraph. They are commonly used in representation theory: a *representation,* V, of a quiver assigns a vector space $V(x)$ to each vertex x of the quiver and a linear map $V(a)$ to each arrow a.

If K is a field and Γ is a quiver, then the quiver algebra or path algebra $K\Gamma$ is defined as follows. Apath in Γ is a sequence of arrows $a_n\, a_{n-1} \ldots a_3\, a_2\, a_1$ such that the head of a_{i+1} = tail of a_i, using the convention of concatenating paths from right to left. Then the path algebra is a vector space having all the paths (of length ≥ 0) in the quiver as basis (including, for each vertex i of the quiver Γ, a *trivial path* e_i of length 0; these paths are *not* assumed to be equal for different i), and multiplication given by concatenation of paths. If two paths cannot be concatenated because the end vertex of the first is not equal to the starting vertex of the second, their product is defined to be zero. This defines an associative algebra over K. This algebra has a unit element if and only if the quiver has only finitely many vertices. In this case, the modules over $K\Gamma$ are naturally identified with the representations of Γ.

If the quiver has finitely many vertices and arrows, and the end vertex and starting vertex of any path are always distinct (i.e. Q has no oriented cycles), then $K\Gamma$ is a finite-dimensional hereditary algebra over K.

Representations of Quivers

A representation V *of a quiver* Q *is said to be* trivial *if* V(x) = 0 *for all vertices* x *in* Q.

A morphism, *f* : V → V', *between representations of the quiver* Q, *is a collection of linear maps* $f(x): V(x) \to V'(x)$ *such that for every arrow* a *in* Q *from* x *to* y V'(a)f(x) = f(y)V(a), *i.e. the squares that f forms with the arrows of* V *and* V' *all commute. A morphism, f, is an* isomorphism, *iff*(x) *is invertible for all vertices* x *in the quiver. With these definitions the representations of a quiver form a category.*

If V and W are representations of a quiver Q, then the direct sum of these representations, $V \oplus W$, is defined by $(V \oplus W)(x) = V(x) \oplus W(x)$ for all vertices x in Q and $(V \oplus W)(a)$ is the direct sum of the linear mappings $V(a)$ and $W(a)$.

A representation is said to be *decomposable* if it is isomorphic to the direct sum of non-zero representations.

A categorical definition of a quiver representation can also be given. The quiver itself can be considered a category, where the vertices are objects and paths are morphisms. Then a representation of Q is just a covariant functor from this category to the category of finite dimensional vector spaces. Morphisms of representations of Q are precisely natural transformations between the corresponding functors.

For a finite quiver Γ (a quiver with finitely many vertices and edges), let $K\Gamma$ be its path algebra. Let e_i denote the trivial path at vertex i. Then we can associate to the vertex i, the projective $K\Gamma$ module $K\Gamma e_i$ consisting of linear combinations of paths which have starting vertex i. This corresponds to the representation of Γ obtained by putting a copy of K at each vertex which lies on a path starting at i and 0 on each other vertex. To each edge joining two copies of K we associate the identity map.

Gabriel's Theorem

A quiver is of finite type if it has only finitely many isomorphism classes of indecomposable representations. Gabriel (1972) classified all quivers of finite type, and also their indecomposable representations. More precisely, Gabriel's theorem states that:

1. A (connected) quiver is of finite type if and only if its underlying graph (when the directions of the arrows are ignored) is one of the ADE Dynkin diagrams: A_n, D_n, E_6, E_7, E_8.
2. The indecomposable representations are in a one-to-one correspondence with the positive roots of the root system of the Dynkin diagram.

Dlab & Ringel (1973) found a generalisation of Gabriel's theorem in which all Dynkin diagrams of finite dimensional semisimple Lie algebras occur.

Term Algebra

In universal algebra and mathematical logic, a term algebra is a freely generated algebraic structure over a given signature. For example, in a signature consisting of a single binary operation, the term algebra over a set X of variables is exactly the free magma generated by X. Other synonyms for the notion include absolutely free algebra, anarchic algebra.

From a category theory perspective, a term algebra is the initial object for the category of all algebras of the same signature, and this object, unique up to isomorphism is called an initial algebra; it generates by homomorphic projection all algebras in the category.

An similar notion is that of a Herbrand universe in logic, usually used under this name in logic programming, which is (absolutely freely) defined starting from the set of constants and function symbols in a set of clauses. That is, the Herbrand universe consists of all ground terms: terms which have no variables in them.

An atomic formula or atom is commonly defined as a predicate applied to a tuple of terms; a ground atom is then a predicate in which only ground terms appear. The Herbrand base is the set of all ground atoms that can be formed from predicate symbols in the original set of clauses and terms in its Herbrand universe. These two concepts are named after Jacques Herbrand.

Term algebras also play a role in the semantics of abstract data types, where an abstract data type declaration provides the signature of a multi-sorted algebraic structure and the term algebra is a concrete model of the abstract declaration.

Decidability of Term Algebras

Term algebras can be shown decidable using quantifier elimination. The complexity of the decision problem is in NONELEMENTARY.

Herbrand Base

The signature σ of a language is a triple <O,F,P> *consisting of the alphabet of constants* O, *the function symbols* F, *and the predicates* P. *The Herbrand base of a signature σ consists of all ground atoms of σ: of all formulas of the form* $R(t_1, \ldots, t_n)$, *where* $t_1, \ldots, t_n$ *are terms containing no variables (i.e. elements of the Herbrand universe) and* R *is an* n-*ary relation symbol (*i.e. *predicate). In the case of logic with equality, it also contains all equations of the form* $t_1=t_2$, *where* t_1 *and* t_2 *contain no variables.*

Algebra Over a Field

In mathematics, an algebra over a field is a vector space equipped with a bilinear vector product. That is to say, it is an algebraic structure consisting of a vector space together with an operation, usually called multiplication, that combines any two vectors to form a third vector; to qualify as an algebra, this multiplication must satisfy certain compatibility axioms with the given vector space structure, such as distributivity. In other words, an algebra over a field is a set together with operations of multiplication, addition, and scalar multiplication by elements of the field.

One may generalise this notion by replacing the field of scalars by a commutative ring, and thus defining an algebra over a ring.

Note that while some authors may use the term "algebra" to describe a unital "associative algebra", this will not be the case for this article.

Definition and Motivation

First Example: The complex Numbers: Any complex number may be written $a + bi$, where a and b are real numbers and i is the imaginary unit. In other words, a complex number is represented by the vector (a, b) over the field of real numbers. So the complex numbers form a two-dimensional real vector space, where addition is given by $(a, b) + (c, d) = (a + c, b + d)$ and scalar multiplication is given by $c(a, b) = (ca, cb)$, where all of a, b, c and d are real numbers. We use the symbol $\cdot$ to multiply two vectors together, which we use complex multiplication to define: $(a, b) \cdot (c, d) = (ac - bd, ad + bc)$.

The following statements are basic properties of the complex numbers. Let x, y, z be complex numbers, and let a, b be real numbers.

- $(x + y) \cdot z = (x \cdot z) + (y \cdot z)$. In other words, multiplying a complex number by the sum of two other complex numbers, is the same as multiplying by each number in the sum, and then adding.
- $(ax) \cdot (by) = (ab)(x \cdot y)$. This shows that complex multiplication is compatible with the scalar multiplication by the real numbers.

This example fits into the following definition by taking the field K to be the real numbers, and the vector space A to be the complex numbers.

Definition

Let K be a field, and let A be a vector space over K equipped with an additional binary operation from $A \times A$ to A, denoted here by $\cdot$ (i.e. if x and y are any two elements of A, $x \cdot y$ is the *product* of x and y). Then A is an algebra over K if the following identities hold for any three elements x, y, and z of A, and all elements ("scalars") a and b of K:

- Left distributivity: $(x + y) \cdot z = x \cdot z + y \cdot z$
- Right distributivity: $x \cdot (y + z) = x \cdot y + x \cdot z$
- Compatibility with scalars: $(ax) \cdot (by) = (ab)(x \cdot y)$.

These three axioms are another way of saying that the binary operation is bilinear. An algebra over $K$is sometimes also called a *K-algebra*, and K is called the *base field* of A. The binary operation is often referred to as *multiplication* in A. The convention adopted in this article is that multiplication of elements of an algebra is not necessarily associative, although some authors use the term *algebra* to refer to an associative algebra.

Notice that when a binary operation on a vector space is commutative, as in the above example of the complex numbers, it is left distributive exactly when it is right distributive. But in general, for non-commutative operations (such as the next example of the quaternions), they are not equivalent, and therefore require separate axioms.

A Motivating Example: Quaternions

The real numbers may be viewed as a *one* dimensional vector space with a compatible multiplication, and hence a one dimensional algebra over itself. We saw above that the complex numbers form a *two dimensional* vector space over the field of real numbers, and hence form a two dimension algebra over the reals. In both these examples, every non-zero vector has an inverse. It is natural to ask whether one can similarly define a multiplication on a *three* dimensional real vector space such that every non-zero element has an inverse. The answer is no.

Nevertheless, in 1843, the quaternions were defined and provided the now famous *four* dimensional example of an algebra over the real numbers, where one can not only multiply vectors, but also divide. Any quaternion may be written as $(a, b, c, d) = a + b\mathrm{i} + c\mathrm{j} + d\mathrm{k}$. Unlike the complex numbers, the quaternions are an example of a non-commutative algebra: for instance, $(0,1,0,0) \cdot (0,0,1,0) = (0,0,0,1)$ but $(0,0,1,0) \cdot (0,1,0,0) = (0,0,0,-1)$.

The quaternions were soon followed by several other hyper complex number systems, which were the early examples of algebras over a field.

Basic Concepts

Algebra Homomorphisms

Given K-algebras A and B, a K-algebra homomorphism is a K-linear map $f: A \to B$ such that $f(\mathrm{xy}) = f(\mathrm{x})\, f(\mathrm{y})$ for all x,y in A. The space of all K-algebra homomorphisms is frequently written as

$$\mathrm{Hom}_{\text{K-alg}}(\mathrm{A},\mathrm{B}).$$

A K-algebra isomorphism is a bijective K-algebra morphism. For all practical purposes, isomorphic algebras differ only by notation.

Subalgebras and Ideals

A *subalgebra* of an algebra over a field K, is a linear subspace that has the property that the product of any two of its elements is again in the subspace. In other words, a subalgebra of an algebra is a subset of elements that is closed under addition, multiplication, and scalar multiplication. In symbols, we say that a subset L of a K-algebra A is a subalgebra if for every x, y in L and c in K, we have that $x \cdot y$, $x + y$, and cx are all in L.

In the above example of the complex numbers viewed as a two-dimensional algebra over the real numbers, the one-dimensional real line is a subalgebra. A *left ideal* of a *K*-algebra is a linear subspace that has the property that any element of the subspace multiplied on the left by any element of the algebra produces an element of the subspace. In symbols, we say that a subset *L* of a *K*-algebra *A* is a left ideal if for every *x* and *y* in *L*, *z* in *A* and *c* in *K*, we have the following three statements.

- 1) $x + y$ is in *L* (*L* is closed under addition),
- 2) cx is in *L* (*L* is closed under scalar multiplication),
- 3) $z \cdot x$ is in *L* (*L* is closed under left multiplication by arbitrary elements).

If (3) were replaced with $x \cdot z$ is in *L*, then this would define a *right ideal*. A *two-sided ideal* is a subset that is both a left and a right ideal. The term *ideal* on its own is usually taken to mean a two-sided ideal. Of course when the algebra is commutative, then all of these notions of ideal are equivalent. Notice that conditions (1) and (2) together are equivalent to *L* being a linear subspace of *A*. It follows from condition (3) that every left or right ideal is a subalgebra.

It is important to notice that this definition is different from the definition of an ideal of a ring, in that here we require the condition (2). Of course if the algebra is unital, then condition (3) implies condition (2).

Extension of Scalars

If we have a field extension *F*/*K*, which is to say a bigger field *F* that contains *K*, then there is a natural way to construct an algebra over *F* from any algebra over *K*. It is the same construction one uses to make a vector space over a bigger field, namely the tensor product $V_F := V \otimes_K F$. So if *A* is an algebra over *K*, then A_F is an algebra over *F*.

Kinds of Algebras and Examples

Algebras over fields come in many different types. These types are specified by insisting on some further axioms, such as commutativity or associativity of the multiplication operation, which are not required in the broad definition of an algebra. The theories corresponding to the different types of algebras are often very different.

Associative Algebras

- the algebra of all *n*-by-*n* matrices over the field (or commutative ring) *K*. Here the multiplication is ordinary matrix multiplication.

- Group algebras, where a group serves as a basis of the vector space and algebra multiplication extends group multiplication.
- the commutative algebra $K[x]$ of all polynomials over K.
- algebras of functions, such as the R-algebra of all real-valued continuous functions defined on the interval [0,1], or the C-algebra of all holomorphic functions defined on some fixed open set in the complex plane. These are also commutative.
- Incidence algebras are built on certain partially ordered sets.
- algebras of linear operators, for example on a Hilbert space. Here the algebra multiplication is given by the composition of operators. These algebras also carry a topology; many of them are defined on an underlying Banach space, which turns them into Banach algebras. If an involution is given as well, we obtain B*-algebras and C*-algebras. These are studied in functional analysis.

Non-associative Algebras

A non-associative algebra (or distributive algebra) over a field K is a K-vector space A equipped with a K-bilinear map $A \times A \to A$. There are left and right multiplication maps $L_a : x \mapsto ax$ and $R_a : x \mapsto xa$. The *enveloping algebra* of A is the subalgebra of all K-endomorphisms of Agenerated by the multiplication maps.

An algebra is *unital* or *unitary* if it has a unit or identity element I with $Ix = x = xI$ for all x in the algebra.

The best-known kinds of non-associative algebras are nearly associative—that is, some simple equation constrains the differences between different ways of associating multiplication of elements. These include:

- Lie algebras that are anticommutative , which require $xx = 0$ and the Jacobi identity $(xy)z + (yz)x + (zx)y = 0$. For these algebras the product is called the *Lie bracket* and is conventionally written $[x,y]$ instead of xy. Examples include:
    - o Euclidean space R^3 with multiplication given by the vector cross product (with K the field R ofr eal numbers);
    - o algebras of vector fields on a differentiable manifold (if K is R or the complex numbers C) or an algebraic variety (for general K);

- every associative algebra gives rise to a Lic algebra by using the commutator as Lie bracket. In fact every Lie algebra can either be constructed this way, or is a subalgebra of a Lie algebra so constructed.

- Jordan algebras that are commutative and satisfy the Jordan property $(xy)x^2 = x(yx^2)$ and also $xy = yx$.
  - every associative algebra over a field of characteristic other than 2 gives rise to a Jordan algebra by defining a new multiplication $x*y = (1/2)(xy + yx)$. In contrast to the Lie algebra case, not every Jordan algebra can be constructed this way. Those that can are called *special*.
- Alternative algebras, which require that $(xx)y = x(xy)$ and $(yx)x = y(xx)$. The most important examples are the octonions (an algebra over the reals), and generalisations of the octonions over other fields. (Obviously all associative algebras are alternative.) Up to isomorphism the only finite-dimensional real alternative, division algebras are the reals, complexes, quaternions and octonions.
- Power-associative algebras, which require that $x^m x^n = x^{m+n}$, where $m \geq 1$ and $n \geq 1$. (Here we formally define x^n recursively as $x(x^{n-1})$.) Examples include all associative algebras, all alternative algebras, and the sedenions.

These properties are related by: associative implies alternative, which in turn implies power associative; commutative and associative implies Jordan, which implies power associative. None of the converse implications hold.

Algebras and Rings

The definition of an associative K-*algebra with unit is also frequently given in an alternative way. In this case, an algebra over a field* K *is a ring* A *together with a ring homomorphism*

$$\eta : K \to Z(A),$$

where $Z(A)$ is the centre of A. Since ç is a ring morphism, then one must have either that A is the trivial ring, or that ç is injective. This definition is equivalent to that above, with scalar multiplication

$$K \times A \to A$$

given by

$$(k,a) \mapsto \eta(k)a.$$

Given two such associative unital K-algebras A and B, a unital K-algebra morphism $f: A \rightarrow B$ is a ring morphism that commutes with the scalar multiplication defined by ç, which one may write as

$$f(ka) = kf(a)$$

for all $k \in K$ and $a \in A$. In other words, the following diagram commutes:

$$\begin{array}{ccccc} & & K & & \\ & \eta_A \swarrow & & \eta_B \searrow & \\ A & & \overset{f}{\rightarrow} & & B \end{array}$$

Structure Coefficients

For algebras over a field, the bilinear multiplication from $A \times A$ to A is completely determined by the multiplication of basis elements of A. Conversely, once a basis for A has been chosen, the products of basis elements can be set arbitrarily, and then extended in a unique way to a bilinear operator on A, i.e., so the resulting multiplication satisfies the algebra laws.

Thus, given the field K, any algebra can be specified up to isomorphism by giving its dimension (say$n$), and specifying n^3 *structure coefficients* $c_{i,j,k}$, which are scalars. These structure coefficients determine the multiplication in A via the following rule:

$$e_i e_j = \sum_{k=1}^{n} c_{i,j,k} e_k$$

where $e_1,...,e_n$ form a basis of A. The only requirement on the structure coefficients is that, if the dimension n is infinite, then this sum must always converge (in whatever sense is appropriate for the situation).

Note however that several different sets of structure coefficients can give rise to isomorphic algebras.

When the algebra can be endowed with a metric, then the structure coefficients are written with upper and lower indices, so as to distinguish their transformation properties under coordinate transformations. Specifically, lower indices are covariant indices, and transform via pullbacks, while upper indices are contravariant, transforming under push forwards. Thus, in mathematical physics, the structure coefficients are often written $c_{i,j}{}^k$, and their defining rule is written using the Einstein notationas

$$e_i e_j = c_{i,j}{}^k e_k.$$

If you apply this to vectors written in index notation, then this becomes

$$(xy)^k = c_{i,j}{}^k x^i y^j.$$

If K is only a commutative ring and not a field, then the same process works if A is a free module over K. If it isn't, then the multiplication is still completely determined by its action on a set that spans A; however, the structure constants can't be specified arbitrarily in this case, and knowing only the structure constants does not specify the algebra up to isomorphism.

Classification of Low-dimensional Algebras

Two-dimensional, three-dimensional and four-dimensional unital associative algebras over the field of complex numbers were completely classified up to isomorphism by Eduard Study.

There exist two two-dimensional algebras. Each algebra consists of linear combinations (with complex coefficients) of two basis elements, 1 (the unit) and a. *According to the definition of a unit,*

$$1 \cdot 1 = 1, \quad 1 \cdot a = a, \quad a \cdot 1 = a.$$

It remains to specify

$aa = 1$ for the first algebra,

$aa = 0$ for the second algebra.

There exist five three-dimensional algebras. Each algebra consists of linear combinations of three basis elements, 1 (the unit), a and b. Taking into account the definition of a unit, it is sufficient to specify

$aa = a, \quad bb = b, \quad ab = ba = 0$ for the first algebra,

$aa = a, \quad bb = 0, \quad ab = ba = 0$ for the second algebra,

$aa = b, \quad bb = 0, \quad ab = ba = 0$ for the third algebra,

$aa = 1, \quad bb = 0, \quad ab = -ba = b$ for the fourth algebra,

$aa = 0, \quad bb = 0, \quad ab = ba = 0$ for the fifth algebra.

The fourth algebra is non-commutative, others are commutative.

Clifford Algebra

In mathematics, Clifford algebras are a type of associative algebra. As K-algebras, they generalise the real numbers, complex numbers, quaternions and several other hyper complex number systems. The theory of Clifford algebras is intimately connected with the theory of quadratic form sand orthogonal transformations. Clifford algebras have important applications in a variety of fields including geometry and theoretical physics. They are named after the English geometre William Kingdon Clifford.

Introduction and Basic Properties

Specifically, a Clifford algebra is a unital associative algebra which contains and is generated by a vector space V *equipped with a quadratic form* Q. *The Clifford algebra* $C\ell(V,Q)$ *is the "freest" algebra generated by* V *subject to the condition*

$$v^2 = Q(v)1 \text{ for all } v \in V.$$

The definition of a Clifford algebra endows it with more structure than a "bare" *K*-algebra, specifically it has a designated or privileged subspace that is isomorphic to *V*. Such a subspace cannot in general be uniquely determined given only a *K*-algebra isomorphic to the Clifford algebra.

If the characteristic of the ground field *K* is not 2, then one can rewrite this fundamental identity in the form

$$uv + vu = 2\langle u, v\rangle \text{ for all } u, v \in V,$$

where $u, v = (Q(u + v) - Q(u) - Q(v))/2$ is the symmetric bilinear form associated with *Q*, via the polarisation identity. The idea of being the "freest" or "most general" algebra subject to this identity can be formally expressed through the notion of a universal property, as done below.

Quadratic forms and Clifford algebras in characteristic 2 form an exceptional case. In particular, if char (K) = 2 it is not true that a quadratic form determines a symmetric bilinear form, or that every quadratic form admits an orthogonal basis. Many of the statements in this article include the condition that the characteristic is not 2, and are false if this condition is removed.

As a Quantization of the Exterior Algebra

Clifford algebras are closely related to exterior algebras. In fact, if $Q = 0$ then the Clifford ℓ algebra C(V,Q) is just the exterior algebra $\Lambda(V)$. For nonzero *Q* there exists a canonical *linear* isomorphism between $\Lambda(V)$ and $C\ell(V,Q)$ whenever the ground field *K* does not have characteristic two. That is, they are naturally isomorphic as vector spaces, but with different multiplications (in the case of characteristic two, they are still isomorphic as vector spaces, just not naturally). Clifford multiplication together with the privileged subspace is strictly richer than the exterior product since it makes use of the extra information provided by *Q*.

More precisely, Clifford algebras may be thought of as *quantizations* (cf. quantization (physics), Quantum group) of the exterior algebra, in the same way that the Weyl algebra is a quantization of the symmetric algebra.

Weyl algebras and Clifford algebras admit a further structure of a *-algebra, and can be unified as even and odd terms of a superalgebra, as discussed in CCR and CAR algebras.

Universal Property and Construction

Let V *be a vector space over a field* K, *and let* $Q : V \to K$ *be a quadratic form on* V. *In most cases of interest the field* K *is either R, C or a finite field.*

A Clifford algebra $C\ell(V,Q)$ *is a unital associative algebra over* K *together with a linear map* $i : V \to C\ell(V,Q)$ *satisfying* $i(v)^2 = Q(v)1$ *for all* $v \in V$, *defined by the following universal property: Given any associative algebra* A *over* K *and any linear map* $j : V \to A$ *such that*

$j(v)^2 = Q(v)1_A$ for all $v \in V$

(where 1_A denotes the multiplicative identity of A), there is a unique algebra homomorphism $f : C\ell(V,Q) \to A$ such that the following diagram commutes (i.e. such that $f \circ i = j$):

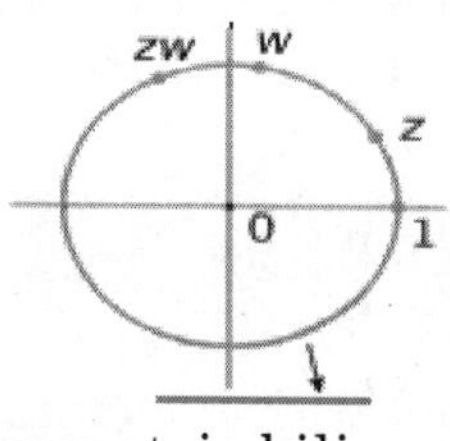

Working with a symmetric bilinear form instead of Q (in characteristic not 2), the requirement on j is

$j(v)j(w) + j(w)j(v) = 2v, w$ for all $v, w \in V$.

A Clifford algebra as described above always exists and can be constructed as follows: start with the most general algebra that contains V, namely the tensor algebra $T(V)$, and then enforce the fundamental identity by taking a suitable quotient. In our case we want to take the two-sided ideal I_Q in $T(V)$ generated by all elements of the form

$v \otimes v - Q(v)1$ for all $v \in V$

and define $C\ell(V,Q)$ as the quotient algebra

$C\ell(V,Q) = T(V)/I_Q$.

The ring product inherited by this quotient is sometimes referred to as the Clifford product to differentiate it from the inner and outer products.

It is then straightforward to show that $C\ell(V,Q)$ contains V and satisfies the above universal property, so that $C\ell$ is unique up to a unique isomorphism; thus one speaks of "the" Clifford algebra $C\ell(V,Q)$. It also follows from this construction that i is injective. One usually drops the i and considers V as a linear subspace of $C\ell(V,Q)$.

The universal characterisation of the Clifford algebra shows that the construction of $C\ell(V,Q)$ is*functorial* in nature. Namely, $C\ell$ can be considered as a functor from the category of vector spaces with quadratic forms (whose morphisms are linear maps preserving the quadratic form) to the category of associative algebras. The universal property guarantees that linear maps between vector spaces (preserving the quadratic form) extend uniquely to algebra homomorphisms between the associated Clifford algebras.

Basis and Dimension

$$\{e_{i_1}e_{i_2}\cdots e_{i_k} \mid 1 \le i_1 < i_2 < \cdots < i_k \le n \text{ and } 0 \le k \le n\}$$

is a basis for $C\ell(V,Q)$. The empty product ($k = 0$) is defined as the multiplicative identity element. For each value of k there are n choose k basis elements, so the total dimension of the Clifford algebra is

$$\dim C\ell(V,Q) = \sum_{k=0}^{n} \binom{n}{k} = 2^n.$$

Since V comes equipped with a quadratic form, there is a set of privileged bases for V: the orthogonalones. An orthogonal basis is one such that

$$\langle e_i, e_j \rangle = 0 \qquad i \ne j.$$

where $\cdot,\cdot$ is the symmetric bilinear form associated to Q. The fundamental Clifford identity implies that for an orthogonal basis

$$e_i e_j = -e_j e_i \qquad i \ne j.$$

This makes manipulation of orthogonal basis vectors quite simple. Given a product $e_{i_1}e_{i_2}\cdots e_{i_k}$ of *distinct* orthogonal basis vectors of V, one can put them into standard order while including an overall sign determined by the number of pairwise swaps needed to do so (i.e. the signature of the ordering permutation).

Examples: Real and Complex Clifford Algebras

The most important Clifford algebras are those over real and complex vector spaces equipped with nondegenerate quadratic forms. The geometric interpretation of nondegenerate real Clifford algebras is known as geometric algebra.

Every nondegenerate quadratic form on a finite-dimensional real vector space is equivalent to the standard diagonal form:

$$Q(v) = v_1^2 + \cdots + v_p^2 - v_{p+1}^2 - \cdots - v_{p+q}^2$$

where $n = p + q$ is the dimension of the vector space. The pair of integers (p, q) is called the signature of the quadratic form. The real vector space with this quadratic form is often denoted $R^{p,q}$. The Clifford algebra on $R^{p,q}$ is denoted $C\ell_{p,q}(R)$. The symbol $C\ell_n(R)$ means either $C\ell_{n,0}(R)$ or $C\ell_{0,n}(R)$ depending on whether the author prefers positive definite or negative definite spaces.

A standard orthonourmal basis $\{e_i\}$ for $R^{p,q}$ consists of $n = p + q$ mutually orthogonal vectors, p of which have norm +1 and q of which have norm -1. The algebra $C\ell_{p,q}(R)$ will therefore have p vectors which square to +1 and q vectors which square to -1.

Note that $C\ell_{0,0}(R)$ is naturally isomorphic to R since there are no nonzero vectors. $C\ell_{0,1}(R)$ is a two-dimensional algebra generated by a single vector e_1 which squares to -1, and therefore is isomorphic to C, the field of complex numbers. The algebra $C\ell_{0,2}(R)$ is a four-dimensional algebra spanned by $\{1, e_1, e_2, e_1e_2\}$. The latter three elements square to -1 and all anticommute, and so the algebra is isomorphic to the quaternions H. The next algebra in the sequence is $C\ell_{0,3}(R)$, and is an 8-dimensional algebra isomorphic to the direct sum $H \oplus H$ called split-biquaternions.

One can also study Clifford algebras on complex vector spaces. Every nondegenerate quadratic form on a complex vector space is equivalent to the standard diagonal form

$$Q(z) = z_1^2 + z_2^2 + \cdots + z_n^2$$

where $n = \dim V$, so there is essentially only one Clifford algebra for each dimension n. We will denote the Clifford algebra on C^n with the standard quadratic form by $C\ell_n(C)$. One can show that the algebra $C\ell_n(C)$ may be obtained as the complexification of the algebra $C\ell_{p,q}(R)$ where $n = p + q$:

$$C\ell_n(\mathbb{C}) \cong C\ell_{p,q}(\mathbb{R}) \otimes \mathbb{C} \cong C\ell(\mathbb{C}^{p+q}, Q \otimes \mathbb{C}).$$

Here Q is the real quadratic form of signature (p,q). Note that the result of the complexification does not depend on the signature. The first few cases are not hard to compute. One finds that

$$C\ell_0(\mathrm{C}) = \mathrm{C}$$
$$C\ell_1(\mathrm{C}) = \mathrm{C} \oplus \mathrm{C}$$
$$C\ell_2(\mathrm{C}) = M_2(\mathrm{C})$$

where $M_2(\mathrm{C})$ denotes the algebra of 2×2 matrices over C.

It turns out that every one of the algebras $C\ell_{p,q}(\mathrm{R})$ and $C\ell_n(\mathrm{C})$ is isomorphic to a matrix algebra over R,C, or H or to a direct sum of two such algebras.

Examples: Quaternions and Dual Quaternions

Quaternions: In this section, Hamilton's quaternions are constructed as the even sub algebra of the Clifford algebra $C\ell_{0,3}(\mathrm{R})$.

Let the vector space V be real three dimensional space R^3, and the quadratic form Q be derived from the usual Euclidean metric. Then, for v, w in R^3 we have the quadratic form, or dot product,

$$\mathrm{v \cdot w} = \mathrm{v_1 w_1 + v_2 w_2 + v_3 w_3}.$$

Now introduce the Clifford product of vectors v and w given by

$$\mathrm{vw + wv} = -2(\mathrm{v \cdot w}).$$

This formulation uses the negative sign so the correspondence with quaternions is easily shown.

Denote a set of orthogonal unit vectors of R^3 as e_1, e_2, and e_3, then the Clifford product yields the relations

$$e_2e_3 = -e_3e_2,\ e_3e_1 = -e_1e_3,\ e_1e_2 = -e_2e_1,$$

and

$$e_1^2 = e_2^2 = e_3^2 = -1.$$

The general element of the Clifford algebra $C\ell_{0,3}(\mathrm{R})$ is given by

$$A = a_0 + a_1e_1 + a_2e_2 + a_3e_3 + a_4e_2e_3 + a_5e_3e_1 + a_6e_1e_2 + a_7e_1e_2e_3.$$

The linear combination of the even rank elements of $C\ell_{0,3}(\mathrm{R})$ defines the even sub algebra $C\ell^0{}_{0,3}(\mathrm{R})$ with the general element

$$Q = q_0 + q_1e_2e_3 + q_2e_3e_1 + q_3e_1e_2.$$

The basis elements can be identified with the quaternion units i, j, k as

$$i = e_2e_3,\ j = e_3e_1,\ k = e_1e_2,$$

which shows that the even sub algebra $C\ell^0_{0,3}(R)$ is Hamilton's real quaternion algebra.

$$i^2 = (e_2e_3)^2 = e_2e_3e_2e_3 = -e_2e_2e_3e_3 = -1,$$

and

$$ij = e_2e_3e_3e_1 = -e_2e_1 = e_1e_2 = k.$$

Finally,

$$ijk = e_2e_3e_3e_1e_1e_2 = -1.$$

Dual Quaternions

In this section, dual quaternions are constructed as the even Clifford algebra of real four dimensional space with a degenerate quadratic form.

Let the vector space V be real four dimensional space R^4, and let the quadratic form Q be a degenerate form derived from the Euclidean metric on R^3. For v, w in R^4 introduce the degenerate bilinear form

$$d(v,w) = v_1w_1 + v_2w_2 + v_3w_3.$$

This degenerate scalar product projects distance measurements in R^4 onto the R^3 hyperplane.

The Clifford product of vectors v and w is given by

$$vw + wv = -2d(v,w).$$

Note the negative sign is introduced to simplify the correspondence with quaternions.

Denote a set of orthogonal unit vectors of R^4 as e_1, e_2, e_3 and e_4, then the Clifford product yields the relations

$$e_me_n = -e_ne_m, m \neq n,$$

and

$$e_1^2 = e_2^2 = e_3^2 = -1, e_4^2 = 0.$$

The general element of the Clifford algebra $C\ell(R^4,d)$ has 16 components. The linear combination of the even ranked elements defines the even sub algebra $C\ell^0(R^4,d)$ with the general element

$$H = h_0 + h_1e_2e_3 + h_2e_3e_1 + h_3e_1e_2 + h_4e_4e_1 + h_5e_4e_2 + h_6e_4e_3 + h_7e_1e_2e_3e_4.$$

The basis elements can be identified with the quaternion units i, j, k and the dual unit ε as

$$i = e_2e_3, j = e_3e_1, k = e_1e_2, \epsilon = e_1e_2e_3e_4.$$

This provides the correspondence of $C\ell^{0}{}_{0,3,1}(\mathrm{R})$ with dual quaternion algebra.

$$\epsilon^2 = (e_1e_2e_3e_4)^2 = e_1e_2e_3e_4e_1e_2e_3e_4 = -e_1e_2e_3(e_4e_4)e_1e_2e_3 = 0,$$

and

$$\epsilon i = (e_1e_2e_3e_4)e_2e_3 = e_1e_2e_3e_4e_2e_3 = e_2e_3(e_1e_2e_3e_4) = i\epsilon.$$

The exchanges of e_1 and e_4 alternate signs an even number of times, and show the dual unit ε commutes with the quaternion units i, j, and k.

Properties

Relation to the Exterior Algebra

Given a vector space V one can construct the exterior algebra $\Lambda(V)$, whose definition is independent of any quadratic form on V. It turns out that if K does not have characteristic 2 then there is a natural isomorphism between $\Lambda(V)$ and $C\ell(V,Q)$ considered as vector spaces (and there exists an isomorphism in characteristic two, which may not be natural). This is an algebra isomorphism if and only if $Q = 0$. One can thus consider the Clifford algebra $C\ell(V,Q)$ as an enrichment (or more precisely, a quantization, cf. the Introduction) of the exterior algebra on V with a multiplication that depends on Q(one can still define the exterior product independent of Q).

The easiest way to establish the isomorphism is to choose an *orthogonal* basis $\{e_i\}$ for V and extend it to a basis for $C\ell(V,Q)$ as described above. The map $C\ell(V,Q)$ ℓ $\Lambda(V)$ is determined by

$$e_{i_1}e_{i_2}\cdots e_{i_k} \mapsto e_{i_1}\wedge e_{i_2}\wedge\cdots\wedge e_{i_k}.$$

Note that this only works if the basis $\{e_i\}$ is orthogonal. One can show that this map is independent of the choice of orthogonal basis and so gives a natural isomorphism.

If the characteristic of K is 0, one can also establish the isomorphism by antisymmetrising. Define functions $f_k : V\times \ldots \times V \to C\ell(V,Q)$ by

$$f_k(v_1,\cdots,v_k) = \frac{1}{k!}\sum_{\sigma\in S_k}\mathrm{sgn}(\sigma)v_{\sigma(1)}\cdots v_{\sigma(k)}$$

where the sum is taken over the symmetric group on k elements. Since f_k is alternating it induces a unique linear map $\Lambda^k(V) \to C\ell(V,Q)$. The direct sum of these maps gives a linear map between $\Lambda(V)$ and $C\ell(V,Q)$. This map can be shown to be a linear isomorphism, and it is natural. A more sophisticated way to view the relationship is to construct a filtration on $C\ell(V,Q)$. Recall that the tensor algebra $T(V)$ has a natural filtration: $F^0 \subset F^1 \subset F^2 \subset \ldots$ where F^k contains sums of tensors with rank $\leq k$. Projecting this down to the Clifford algebra gives a filtration on $C\ell(V,Q)$. The associated graded algebra

$$Gr_F C\ell(V,Q) = \bigoplus_k F^k / F^{k-1}$$

is naturally isomorphic to the exterior algebra $\Lambda(V)$. Since the associated graded algebra of a filtered algebra is always isomorphic to the filtered algebra as filtered vector spaces (by choosing complements of F^k in F^{k+1} for all k), this provides an isomorphism (although not a natural one) in any characteristic, even two.

Grading

In the following, assume that the characteristic is not 2.

Clifford algebras are Z_2-graded algebras (also known as superalgebras). Indeed, the linear map on Vdefined by $v \mapsto -v$ (reflection through the origin) preserves the quadratic form Q and so by the universal property of Clifford algebras extends to an algebra automorphism

$$\alpha : C\ell(V,Q) \to C\ell(V,Q).$$

Since α is an involution (i.e. it squares to the identity) one can decompose $C\ell(V,Q)$ into positive and negative eigenspaces of α

$$C\ell(V,Q) = C\ell^0(V,Q) \oplus C\ell^1(V,Q)$$

where $C\ell^i(V,Q) = \{x \in C\ell(V,Q) \mid \alpha(x) = (-1)^i x\}$. Since α is an automorphism it follows that

$$C\ell^i(V,Q)C\ell^j(V,Q) = C\ell^{i+j}(V,Q)$$

where the superscripts are read modulo 2. This gives $C\ell(V,Q)$ the structure of a Z_2-graded algebra. The subspace $C\ell^0(V,Q)$ forms a subalgebra of $C\ell(V,Q)$, called the *even subalgebra*. The subspace$C\ell^1(V,Q)$ is called the *odd part* of $C\ell(V,Q)$ (it is not a subalgebra). This Z_2-grading plays an important role in the analysis and application of Clifford algebras. The automorphism α is called the *maininvolution* or *grade involution*. Elements that are pure in this Z_2-grading are simply said to be even or odd.

Remark. In characteristic not 2 the underlying vector space of $C\ell(V,Q)$ inherits an N-grading and a Z-grading from the canonical isomorphism with the underlying vector space of the exterior algebra

$\Lambda(V)$. It is important to note, however, that this is a *vector space grading only*. That is, Clifford multiplication does not respect the N-grading or Z-grading, only the Z_2-grading: for instance if $Q(v) \neq 0$, then $v \in C\ell^1(V,Q)$, but $v^2 \in C\ell^0(V,Q)$, not in $C\ell^2(V,Q)$. Happily, the gradings are related in the natural way: $Z_2 \cong N/2N \cong Z/2Z$. Further, the Clifford algebra is Z-filtered: $C\ell^{\leq i}(V,Q) \cdot C\ell^{\leq j}\}(V,Q) \subset C\ell^{\leq i+j}(V,Q)$. The *degree* of a Clifford number usually refers to the degree in the N-grading.

The even subalgebra $C\ell^0(V,Q)$ of a Clifford algebra is itself isomorphic to a Clifford algebra. If V is the orthogonal direct sum of a vector a of norm $Q(a)$ and a subspace U, then $C\ell^0(V,Q)$ is isomorphic to $C\ell(U,-Q(a)Q)$, where $-Q(a)Q$ is the form Q restricted to U and multiplied by $-Q(a)$. In particular over the reals this implies that

$$C\ell^0_{p,q}(\mathbb{R}) \cong C\ell_{p,q-1}(\mathbb{R}) \text{ for } q > 0, \text{ and}$$

$$C\ell^0_{p,q}(\mathbb{R}) \cong C\ell_{q,p-1}(\mathbb{R}) \text{ for } p > 0.$$

In the negative-definite case this gives an inclusion $C\ell_{0,n-1}(R) \subset C\ell_{0,n}(R)$ which extends the sequence

$$R \subset C \subset H \subset H \oplus H \subset \ldots$$

Likewise, in the complex case, one can show that the even subalgebra of $C\ell_n(C)$ is isomorphic to $C\ell_{n-1}(C)$.

Antiautomorphisms

In addition to the automorphism α, there are two antiautomorphisms which play an important role in the analysis of Clifford algebras. Recall that the tensor algebra $T(V)$ comes with an antiautomorphism that reverses the order in all products:

$$v_1 \otimes v_2 \otimes \cdots \otimes v_k \mapsto v_k \otimes \cdots \otimes v_2 \otimes v_1.$$

Since the ideal I_Q is invariant under this reversal, this operation descends to an antiautomorphism ℓ of $C(V,Q)$ called the *transpose* or *reversal* operation, denoted by x^t. The transpose is an antiautomorphism: $(xy)^t = y^t x^t$. The transpose operation makes no use of the Z_2-grading so we define a second antiautomorphism by composing α and the transpose. We call this operation *Clifford conjugation* denoted $\bar{x}$

$$\bar{x} = \alpha(x^t) = \alpha(x)^t.$$

Of the two antiautomorphisms, the transpose is the more fundamental.

Note that all of these operations are involutions. One can show that they act as ±1 on elements which are pure in the Z-grading. In fact, all three operations depend only on the degree modulo 4. That is, if *x* is pure with degree *k* then

$$\alpha(x) = \pm x \qquad x^t = \pm x \qquad \bar{x} = \pm x$$

where the signs are given by the following table:

k mod 4	0	1	2	3	
	+	-	+	-	$(-1)^k$
	+	+	-	-	$(-1)^{k(k-1)/2}$
	+	-	-	+	$(-1)^{k(k+1)/2}$

The Clifford Scalar Product

When the characteristic is not 2, the quadratic form *Q* on *V* can be extended to a quadratic form on all of $C\ell(V,Q)$ (which we also denoted by *Q*). A basis independent definition of one such extension is

$$Q(x) = \langle x^t x \rangle$$

where *a* denotes the scalar part of *a* (the grade 0 part in the Z-grading). One can show that

$$Q(v_1 v_2 \cdots v_k) = Q(v_1)Q(v_2)\cdots Q(v_k)$$

where the v_i are elements of *V* – this identity is *not* true for arbitrary elements of $C\ell(V,Q)$.

The associated symmetric bilinear form on $C\ell(V,Q)$ is given by

$$\langle x,y \rangle = \langle x^t y \rangle.$$

One can check that this reduces to the original bilinear form when restricted to *V*. The bilinear form on all of $C\ell(V,Q)$ is nondegenerate if and only if it is nondegenerate on *V*.

It is not hard to verify that the transpose is the adjoint of left/right Clifford multiplication with respect to this inner product. That is,

$$\langle ax,y \rangle = \langle x,a^t y \rangle, \text{ and}$$

$$\langle xa,y \rangle = \langle x,ya^t \rangle.$$

Structure of Clifford Algebras

In this section we assume that the vector space *V* is finite dimensional and that the bilinear form of *Q* is non-singular. A central simple algebra over *K* is a matrix algebra over a (finite dimensional) division algebra with centre *K*. For example, the central simple algebras over the reals are matrix algebras over either the reals or the quaternions.

- If *V* has even dimension then $C\ell(V,Q)$ is a central simple algebra over *K*.
- If *V* has even dimension then $C\ell^0(V,Q)$ is a central simple algebra over a quadratic extension of *K* or a sum of two isomorphic central simple algebras over *K*.
- If *V* has odd dimension then $C\ell(V,Q)$ is a central simple algebra over a quadratic extension of *K* or a sum of two isomorphic central simple algebras over *K*.
- If *V* has odd dimension then $C\ell^0(V,Q)$ is a central simple algebra over *K*.

The structure of Clifford algebras can be worked out explicitly using the following result. Suppose that U has even dimension and a non-singular bilinear form with discriminant *d*, and suppose that *V* is another vector space with a quadratic form.

The Clifford algebra of *U*+*V* is isomorphic to the tensor product of the Clifford algebras of *U* and $(-1)^{\dim(U)/2}dV$, which is the space *V* with its quadratic form multiplied by $(-1)^{\dim(U)/2}d$. Over the reals, this implies in particular that

$$C\ell_{p+2,q}(\mathbb{R}) = M_2(\mathbb{R}) \otimes C\ell_{q,p}(\mathbb{R})$$

$$C\ell_{p+1,q+1}(\mathbb{R}) = M_2(\mathbb{R}) \otimes C\ell_{p,q}(\mathbb{R})$$

$$C\ell_{p,q+2}(\mathbb{R}) = \mathbb{H} \otimes C\ell_{q,p}(\mathbb{R}).$$

These formulas can be used to find the structure of all real Clifford algebras and all complex Clifford algebras.

Notably, the Morita equivalence class of a Clifford algebra (its representation theory: the equivalence class of the category of modules over it) depends only on the signature $(p - q)$ mod 8. This is an algebraic form of Bott periodicity.

The Clifford Group Γ

In this section we assume that *V* is finite dimensional and the quadratic form *Q* is nondegenerate.

The invertible elements of the Clifford algebra act on it by twisted conjugation: conjugation by *x* maps $xv\alpha(x)^{-1} \in V$.

The Clifford group Γ is defined to be the set of invertible elements x that *stabilise vectors*, meaning that

$$1 \to K^* \to \Gamma \to O_V(K) \to 1,$$

for all v in V.

This formula also defines an action of the Clifford group on the vector space V that preserves the norm Q, and so gives a homomorphism from the Clifford group to the orthogonal group. The Clifford group contains all elements r of V of nonzero norm, and these act on V by the corresponding reflections that take v to v - $2v,rr/Q(r)$ (In characteristic 2 these are called orthogonal transvections rather than reflections.)

The Clifford group Γ is the disjoint union of two subsets Γ^0 and Γ^1, where Γ^i is the subset of elements of degree i. The subset Γ^0 is a subgroup of index 2 in Γ.

If V is a finite dimensional real vector space with positive definite (or negative definite) quadratic form then the Clifford group maps onto the orthogonal group of V with respect to the form (by the Cartan-Dieudonné theorem) and the kernel consists of the nonzero elements of the field K. This leads to exact sequences

$$1 \to K^* \to \Gamma \to O_V(K) \to 1,$$

$$1 \to K^* \to \Gamma^0 \to SO_V(K) \to 1.$$

Over other fields or with indefinite forms, the map is not in general onto, and the failure is captured by the spinor norm.

Spinor Norm

In arbitrary characteristic, the spinor norm Q is defined on the Clifford group by

$$Q(x) = x^t x.$$

It is a homomorphism from the Clifford group to the group K^* of non-zero elements of K. It coincides with the quadratic form Q of V when V is identified with a subspace of the Clifford algebra. Several authors define the spinor norm slightly differently, so that it differs from the one here by a factor of -1, 2, or -2 on Γ^1. The difference is not very important in characteristic other than 2.

The nonzero elements of K have spinor norm in the group K^{*2} of squares of nonzero elements of the field K. So when V is finite dimensional and non-singular we get an induced map from the orthogonal group of V to the group K^*/K^{*2}, also called the spinor norm. The spinor norm of the reflection of a vector r has image $Q(r)$ in K^*/K^{*2}, and this property uniquely defines it on the orthogonal group. This gives exact sequences:

$$1 \to \{\pm 1\} \to \mathrm{Pin}_V(K) \to O_V(K) \to K^* / K^{*2},$$

$$1 \to \{\pm 1\} \to \mathrm{Spin}_V(K) \to SO_V(K) \to K^* / K^{*2}.$$

Note that in characteristic 2 the group $\{\pm 1\}$ has just one element.

From the point of view of Galois cohomology of algebraic groups, the spinor norm is a connecting homomorphism on cohomology. Writing μ_2 for the algebraic group of square roots of 1 (over a field of characteristic not 2 it is roughly the same as a two-element group with trivial Galois action), the short exact sequence

$$1 \to \mu_2 \to \mathrm{Pin}_V \to O_V \to 1$$

yields a long exact sequence on cohomology, which begins

$$1 \to H^0(\mu_2;K) \to H^0(\mathrm{Pin}_V;K) \to H^0(O_V;K) \to H^1(\mu_2;K).$$

The 0th Galois cohomology group of an algebraic group with coefficients in K is just the group of K-valued points: $H^0(G;K) = G(K)$, and $H^1(\mu_2;K) \cong K^* / K^{*2}$, which recovers the previous sequence

$$1 \to \{\pm 1\} \to \mathrm{Pin}_V(K) \to O_V(K) \to K^* / K^{*2},$$

where the spinor norm is the connecting homomorphism $H^0(O_V;K) \to H^1(\mu_2;K)$.

Spin and Pin Groups

In this section we assume that V is finite dimensional and its bilinear form is non-singular. (If K has characteristic 2 this implies that the dimension of V is even.)

The Pin group $\mathrm{Pin}_V(K)$ is the subgroup of the Clifford group Γ of elements of spinor norm 1, and similarly the Spin group $\mathrm{Spin}_V(K)$ is the subgroup of elements of Dickson invariant 0 in $\mathrm{Pin}_V(K)$. When the characteristic is not 2, these are the elements of determinant 1. The Spin group usually has index 2 in the Pin group.

Recall from the previous section that there is a homomorphism from the Clifford group onto the orthogonal group. We define the special orthogonal group to be the image of Γ^0. If K does not have characteristic 2 this is just the group of elements of the orthogonal group of determinant 1. If K does have characteristic 2, then all elements of the orthogonal group have determinant 1, and the special orthogonal group is the set of elements of Dickson invariant 0.

There is a homomorphism from the Pin group to the orthogonal group. The image consists of the elements of spinor norm 1 - K^*/K^{*2}. The kernel consists of the elements +1 and -1, and has order 2 unless *K* has characteristic 2. Similarly there is a homomorphism from the Spin group to the special orthogonal group of *V*.

In the common case when *V* is a positive or negative definite space over the reals, the spin group maps onto the special orthogonal group, and is simply connected when *V* has dimension at least 3. Further the kernel of this homomorphism consists of 1 and -1. So in this case the spin group, Spin(*n*), is a double cover of SO(*n*).

Please note, however, that the simple connectedness of the spin group is not true in general: if *V* is $R^{p,q}$ for *p* and *q* both at least 2 then the spin group is not simply connected. In this case the algebraic group $\mathrm{Spin}_{p,q}$ is simply connected as an algebraic group, even though its group of real valued points $\mathrm{Spin}_{p,q}(R)$ is not simply connected. This is a rather subtle point, which completely confused the authors of at least one standard book about spin groups.

Spinors

Clifford algebras $C\ell_{p,q}(C)$, *with* p+q=2n *even, are matrix algebras which have a complex representation of dimension* 2^n. *By restricting to the group* $Pin_{p,q}(R)$ *we get a complex representation of the Pin group of the same dimension, called the spin representation. If we restrict this to the spin group* $Spin_{p,q}(R)$ *then it splits as the sum of two* half spin representations *(or* Weyl representations*) of dimension* 2^{n-1}.

If p+q=2n+*1 is odd then the Clifford algebra* $C\ell_{p,q}(C)$ *is a sum of two matrix algebras, each of which has a representation of dimension* 2^n, *and these are also both representations of the Pin group* $Pin_{p,q}(R)$. *On restriction to the spin group* $Spin_{p,q}(R)$ *these become isomorphic, so the spin group has a complex spinor representation of dimension* 2^n.

More generally, spinor groups and pin groups over any field have similar representations whose exact structure depends on the structure of the corresponding Clifford algebras: whenever a Clifford algebra has a factor that is a matrix algebra over some division algebra, we get a corresponding representation of the pin and spin groups over that division algebra.

Real Spinors

To describe the real spin representations, one must know how the spin group sits inside its Clifford algebra. The Pin group, $\mathrm{Pin}_{p,q}$ is the set of invertible elements in $C\ell_{p,q}$ which can be written as a product of unit vectors:

$$\mathrm{Pin}_{p,q} = \{v_1 v_2 \ldots v_r \mid \forall i, \mid |v_i| \models \pm 1\}.$$

Comparing with the above concrete realisations of the Clifford algebras, the Pin group corresponds to the products of arbitrarily many reflections: it is a cover of the full orthogonal group $O(p,q)$. The Spin group consists of those elements of $\mathrm{Pin}_{p,q}$ which are products of an even number of unit vectors. Thus by the Cartan-Dieudonné theorem Spin is a cover of the group of proper rotations $SO(p,q)$.

Let $\alpha: C\ell \to C\ell$ be the automorphism which is given by the mapping $v \to -v$ acting on pure vectors. Then in particular, $\mathrm{Spin}_{p,q}$ is the subgroup of $\mathrm{Pin}_{p,q}$ whose elements are fixed by α. Let

$$C\ell^0_{p,q} = \{x \in C\ell_{p,q} \mid \alpha(x) = x\}.$$

(These are precisely the elements of even degree in $C\ell_{p,q}$.) Then the spin group lies within $C\ell^0_{p,q}$.

The irreducible representations of $C\ell_{p,q}$ restrict to give representations of the pin group. Conversely, since the pin group is generated by unit vectors, all of its irreducible representation are induced in this manner. Thus the two representations coincide. For the same reasons, the irreducible representations of the spin coincide with the irreducible representations of $C\ell^0_{p,q}$

To classify the pin representations, one need only appeal to the classification of Clifford algebras. To find the spin representations (which are representations of the even subalgebra), one can first make use of either of the isomorphisms.

$$C\ell^0_{p,q} \approx C\ell_{p,q-1}, \text{ for } q > 0$$
$$C\ell^0_{p,q} \approx C\ell_{q,p-1}, \text{ for } p > 0$$

and realise a spin representation in signature (p,q) as a pin representation in either signature $(p,q\text{-}1)$ or $(q,p\text{-}1)$.

Applications

Differential Geometry: One of the principal applications of the exterior algebra is in differential geometry where it is used to define the bundle of differential forms on a smooth manifold. In the case of a (pseudo-) Riemannian manifold, the tangent spaces come equipped with a natural quadratic form induced by the metric. Thus, one can define a Clifford bundle in analogy with the exterior bundle. This has a number of important applications in Riemannian geometry. Perhaps more importantly is the link to a spin manifold, its associated spinor bundle and spin^c manifolds.

Physics: Clifford algebras have numerous important applications in physics. Physicists usually consider a Clifford algebra to be an algebra spanned by matrices $\gamma_0,\ldots,\gamma_3$ called Dirac matrices which have the property that

$$\gamma_i\gamma_j + \gamma_j\gamma_i = 2\eta_{ij}$$

where η is the matrix of a quadratic form of signature (1,3). These are exactly the defining relations for the Clifford algebra $C\ell_{1,3}(\mathbb{C})$ (up to an unimportant factor of 2), which by the classification of Clifford algebras is isomorphic to the algebra of 4 by 4 complex matrices.

The Dirac matrices were first written down by Paul Dirac when he was trying to write a relativistic first-order wave equation for the electron, and give an explicit isomorphism from the Clifford algebra to the algebra of complex matrices. The result was used to define the Dirac equation and introduce the Dirac operator. The entire Clifford algebra shows up in quantum field theory in the form of Dirac field bilinears.

Computer Vision: Recently, Clifford algebras have been applied in the problem of action recognition and classification in computer vision. Rodriguez et al. propose a Clifford embedding to generalise traditional MACH ûlters to video (3D spatiotemporal volume), and vector-valued data such as optical flow. Vector-valued data is analysed using the Clifford Fourier transform. Based on these vectors action filters are synthesized in the Clifford Fourier domain and recognition of actions is performed using Clifford Correlation. The authors demonstrate the effectiveness of the Clifford embedding by recognising actions typically performed in classic feature ûlms and sports broadcast television.

Chapter 3

Differential Algebra

In mathematics, differential rings, differential fields, and differential algebras are rings, fields, and algebras equipped with a derivation, which is a unary function that is linear and satisfies the Leibniz product law. A natural example of a differential field is the field of rational functions *C*(*t*) in one variable, over the complex numbers, where the derivation is differentiation with respect to *t*.

Differential Ring

A differential ring is a ring R *equipped with one or more derivations, that is additive homomorphisms*

$$\partial : \mathrm{R} \rightarrow \mathrm{R}$$

such that each derivation satisfies the Leibniz product rule

$$\partial(\mathrm{r}_1\mathrm{r}_2) = (\partial\mathrm{r}_1)\mathrm{r}_2 + \mathrm{r}_1(\partial\mathrm{r}_2),$$

for every $\mathrm{r}_1, \mathrm{r}_2 \in \mathrm{R}$. Note that the ring could be noncommutative, so the somewhat standard $d(xy) = xdy + ydx$ form of the product rule in commutative settings may be false. If $\mathrm{M} : \mathrm{R} \times \mathrm{R} \rightarrow \mathrm{R}$ is multiplication on the ring, the product rule is the identity

$$\partial^\circ\mathrm{M} = \mathrm{M}^\circ(\partial \otimes \mathrm{id}) + \mathrm{M}^\circ(\mathrm{id} \otimes \partial).$$

where $\mathrm{f} \otimes \mathrm{g}$ means the function which maps a pair (x,y) to the pair $(f(x),g(y))$.

Differential Field

A differential field is a field *K*, together with a derivation. The theory of differential fields, *DF*, is given by the usual field axioms along with two extra axioms involving the derivation. As above, the derivation must obey the product rule, or Leibniz rule over the elements of the field, to be worthy of being called a derivation. That is, for any two elements *u*, *v* of the field, one has

$$\partial(uv) = u\partial v + v\partial u$$

since multiplication on the field is commutative. The derivation must also be distributive over addition in the field:

$$\partial(u + v) = \partial u + \partial v\ .$$

If *K* is a differential field then *the field of constants* $k = \{u \in K : \partial(u) = 0\}$.

Differential Algebra

A differential algebra over a field K *is a* K*-algebra* A *wherein the derivation(s) commutes with the field. That is, for all* $k \in K$ *and* $x \in A$ *one has*

$$\partial(kx) = k\partial x$$

In index-free notation, if $\eta : K \to A$ is the ring morphism defining scalar multiplication on the algebra, one has

$$\partial^\circ M^\circ(\eta \times Id) = M^\circ(\eta \times \partial)$$

As above, the derivation must obey the Leibniz rule over the algebra multiplication, and must be linear over addition. Thus, for all $a,b \in K$ and $x,y \in A$ one has

$$\partial(xy) = (\partial x)y + x(\partial y)$$

and

$$\partial(ax + by) = a\partial x + b\partial y.$$

Derivation on a Lie Algebra

A derivation on a Lie algebra $\mathfrak{g}$ is a linear map $D : \mathfrak{g} \to \mathfrak{g}$ satisfying the Leibniz rule:

$$D([a,b]) = [a,D(b)] + [D(a),b]$$

For any $a \in \mathfrak{g}$, ad(*a*) is a derivation on $\mathfrak{g}$, which follows from the Jacobi identity. Any such derivation is called an inner derivation.

Examples

If A *is unital, then* $\partial(1) = 0$ *since* $\partial(1) = \partial(1 \times 1) = \partial(1) + \partial(1)$. *For example, in a differential field of characteristic zero the rationals are always a subfield of the constant field.*

Any field pure can be interpreted as a constant differential field.

The field $Q(t)$ *has a unique structure as a differential field, determined by setting* $\partial(t) = 1$*: the field axioms along with the axioms for derivations ensure that the derivation is differentiation with respect tot. For example, by commutativity of multiplication and the Leibniz* *law one has that* $\partial(u^2) = u\,\partial(u) + \partial(u)u = 2u\partial(u)$. *The differential field* $Q(t)$ *fails to have a solution to the differential equation*

$$\partial(u) = u$$

but expands to a larger differential field including the function e^t which does have a solution to this equation. A differential field with solutions to all systems of differential equations is called a differentially closed field. Such fields exist, although they do not appear as natural algebraic or geometric objects. All differential fields (of bounded cardinality) embed into a large differentially closed field. Differential fields are the objects of study in differential Galois theory.

Naturally occurring examples of derivations are partial derivatives, Lie derivatives, the Pincherle derivative, and the commutator with respect to an element of the algebra. All these examples are tightly related, with the concept of derivation as the major unifying theme.

Ring of Pseudo-differential Operators

Differential rings and differential algebras are often studied by means of the ring of pseudo-differential operators on them.

This is the ring

$$R((\xi^{-1})) = \left\{ \sum_{n<\infty} r_n \xi^n \mid r_n \in R \right\}.$$

Multiplication on this ring is defined as

$$(r\xi^m)(s\xi^n) = \sum_{k=0}^{m} r(\partial^k s)\binom{m}{k}\xi^{m+n-k}.$$

Here $\binom{m}{k}$ is the binomial coefficient. Note the identities

$$\xi^{-1}r = \sum_{n=0}^{\infty} (-1)^n (\partial^n r)\xi^{-1-n}$$

which makes use of the identity

$$\binom{-1}{n} = (-1)^n$$

and

$$r\xi^{-1} = \sum_{n=0}^{\infty} \xi^{-1-n}(\partial^n r).$$

Geometric Algebra

Geometric algebra (GA), together with the associated Geometric calculus, provides a comprehensive alternative approach to the algebraic representation of classical, computational and relativistic geometry. GA now finds application in all of physics, in graphics and in robotics. GA is Clifford algebra over the field of real numbers, where the naturally embedded Grassmann algebra plays a part. Spacetime algebra and Conformal Geometric algebra are specific examples of GA.

A feature of GA is its geometric interpretation due to the natural correspondence between geometric entities and the elements of the algebra. GA allows direct manipulation of subspaces and is a coordinate-free formalism. Proponents argue that it provides compact and intuitive descriptions in many areas including classical and quantum mechanics, electromagnetic theory and relativity among others. In many cases, the "proliferation of manifolds" that occurs during the standard curriculum of differential geometry can be reduced by working with real-only algebras.

Definition and Notation

This section begins with a brief, formal description, but then it will demonstrate easy to understand properties of the geometric algebra in a concrete way. Given a finite dimensional real quadratic space $V = \mathrm{R}^n$ with quadratic form $Q : V \to \mathrm{R}$, the geometric algebra for this quadratic space is the Clifford algebra $C\ell(V,Q)$.

The algebra product is called the geometric product, and it is just the Clifford product. It is standard to denote the geometric product by juxtaposition.

For bilinear forms of any signature, an orthogonal basis $\{e_1,...,e_n\}$ can be found for V such that each e_i^2 is either -1, 0 or +1. The number of each case is dictated by the signature.

When Q is nondegenerate there are no 0's in the signature, and so an orthogonal basis of V exists with p elements squaring to 1 and q elements squaring to -1, with $p + q = n$. We denote this algebra $\mathcal{G}(\mathrm{p,q})$. For example, $\mathcal{G}(3,0)$ models 3D Euclidean space, $\mathcal{G}(1,3)$ relativistic spacetime and $\mathcal{G}(4,1)$ a 3D Conformal Geometric algebra.

Standard Bases and Grading

The geometric product creates a symmetric bilinear form $\cdot : V \times V \to \mathrm{R}$ given by $a\,.\,b = (ab + ba)/2$. This is the *inner product* for V discussed below. With respect to this bilinear form, a basis $\{e_1,\cdots,e_n\}$ for V can be found such that

$$e_i \cdot e_j = 0 \text{ for all } i \neq j \text{ (othogonality) and}$$

$$e_i \cdot e_i = e_i e_i = Q(e_i) \in \{-1,0,1\}$$

The set of all possible products of these n symbols with indices in increasing order, including 1 as the empty product, forms a basis for the geometric algebra. As an illustration, the following is a basis for the geometric algebra $\mathcal{G}(3,0)$:

$$\{1, e_1, e_2, e_3, e_1e_2, e_1e_3, e_2e_3, e_1e_2e_3\}$$

A basis formed this way is called a standard basis for the geometric algebra, and any other orthogonal basis for V fitting the above description will produce another standard basis. Each standard basis consists of 2^n elements. The geometric product between elements of the algebra is completely described by the rules:

- $e_i e_i = Q(e_i)$
- $e_i e_j = -e_j e_i$ for $i \neq j$ (orthogonal vectors anticommute)
- $e_i \alpha = \alpha e_i$ (scalars commute)
- $(AB)C = A(BC) = ABC$ (associativity of the geometric product)
- $(A+B)C = AC + BC$ and $C(A+B) = CA + CB$ (distributivity of the geometric product over addition).

These first four rules give the geometric product of any two elements in the standard basis as another up to a sign or zero. A few example computations follow:

$$(e_1e_2)(e_3e_4) = e_1e_2e_3e_4$$

$$(e_2)(e_1e_2e_4) = (e_2e_1)(e_2e_4) = -(e_1e_2)(e_2e_4) = -e_1(e_2e_2)e_4 = -Q(e_2)e_1e_4.$$

The geometric product of any two elements in the algebra can be computed with these rules, including the last. Specifically, if the standard basis elements $\{b_i \mid i \in S\}$ with S being an index set, then

$$(\Sigma_i \alpha_i b_i)(\Sigma_j \beta_j b_j) = \Sigma_{i,j} \alpha_i \beta_j b_i b_j$$

Standard bases may be used to define grades (not to be confused with a grading over the geometric product) of some elements in the geometric algebra. Normal vectors in the span of $\{e_1,\cdots,e_n\}$ are called 1-vectors. Scalars are called 0-vectors. Elements in the span of $\{e_ie_j \mid 1 \leq i < j \leq n\}$ are called 2-vectors, and elements in the span of $\{e_i e_j e_k \mid 1 \le i < j < k \le n\}$ are called 3-vectors and so on until the last grade of *n*-vectors. Many of the elements of the algebra are not graded by this scheme since they are sums of elements of differing grade. Such elements are said to be of *mixed grade.*

Generic elements of the geometric algebra are usually called multivectors with the term *vector usually* reserved for 1-vectors. The grading of multivectors is independent of the orthogonal basis chosen originally.

A multivector *A* may be decomposed with the grade projection operator $\langle A \rangle_r$ which outputs the grade *r* portion of *A*. As a result:

$$A = \sum_{r=0}^{n} \langle A \rangle_r$$

As an example, the geometric product of two vectors $ab = a \cdot b + a \wedge b = \langle ab \rangle_0 + \langle ab \rangle_2$ since $\langle ab \rangle_0 = a \cdot b$ and $\langle ab \rangle_2 = a \wedge b$ and $\langle ab \rangle_i = 0$ for *i* other than 0 and 2.

Inner and Outer Products

There are two other important operations on $V \times V$ in the geometric algebra besides the geometric product. Let *a* and *b* be elements of *V*:

- The inner product on *V* is the symmetric bilinear form arising as the symmetric part of geometric multiplication (and equivalently the bilinear form arising from the quadratic form *Q*), and is denoted by $a \cdot b$. It is related to the geometric product and quadratic form by these equations:

$$a \cdot b = \frac{1}{2}(ab + ba) = b \cdot a$$

$$aa = a \cdot a = Q(a)$$

The inner product of two vectors is always a 0-vector of the algebra.

- The outer product on *V*, denoted with Ω, arises as the antisymmetric part of the geometric product:

$$a \wedge b = \frac{1}{2}(ab - ba) = -b \wedge a$$

$$a \wedge a = 0$$

The outer product of two vectors is always a 2-vector of the algebra.

- The inner and outer products are united into the geometric product since

$$ab = a \cdot b + a \wedge b,$$

Thus the geometric product of two vectors is in general of mixed grade.

If $a.a \neq 0$ for some vector a, then a^{-1} exists and is equal to $a/(a.a)$. For a positive-definite or negative-definite quadratic form, all nonzero vectors have multiplicative inverses. Not all the elements of the algebra are invertible. For example, if u is a unit vector in V (i.e. a vector such that $uu = 1$), the elements $1 \pm u$ have no inverse since they are zero divisors: $(1 - u)(1 + u) = 1 - uu = 1 - 1 = 0$.

Vectors will be represented by lower case letters (e.g. a), and multivectors by upper case letters (e.g.A). Scalars will be represented by Greek characters.

Representation of Subspaces

Geometric algebra represents subspaces of V as multivectors, and so they coexist in the same algebra with vectors from V. A k dimensional subspace W of V is represented by taking an orthogonal basis $\{b_1, b_2, \cdots b_k\}$ and using the geometric product to form the blade $D = b_1 b_2 ... b_k$. There are multiple blades representing W; all those representing W are scalar multiples of D. These blades can be separated into two sets: positive multiples of D and negative multiples of D. The positive multiples of D are said to have *the same orientation* as D, and the negative multiples the *opposite orientation*.

Blades are important since geometric operations such as projections, rotations and reflections are implemented by using the geometric product to multiply vectors and blades.

Unit Pseudoscalars

Pseudoscalars are blades that play important roles in GA. A pseudoscalar for a subspace W of V is a blade that is the product of the members of an orthonourmal basis for W. It can be shown that if I and$I2$ are both pseudoscalars for W, then $I = \pm I2$.

Suppose the geometric algebra $\mathcal{G}(n,0)$ with the familiar positive definite inner product on R^n is formed. Given a plane (2-dimensional subspace) of R^n, one can find an orthonourmal basis $\{b_1, b_2\}$ spanning the plane, and thus find a pseudoscalar $I = b_1 b_2$ representing this plane. The geometric product of any two vectors in the span of b_1 and b_2 lies in $\{\alpha_0 + \alpha_1 I \mid \alpha_i \in \mathbb{R}\}$, that is, it is the sum of a 0-vector and a 2-vector.

By the properties of the geometric product, $I^2 = b_1b_2b_1b_2 = -b_1b_2b_2b_1 = -1$. The resemblance to the imaginary unit is not accidental: the subspace $\{\alpha_0 + \alpha_1 I \mid \alpha_i \in \mathbb{R}\}$ is R-algebra isomorphic to the complex numbers. In this way, a copy of the complex numbers is embedded in the geometric algebra for each 2-dimensional subspace of V.

It is sometimes possible to identify the presence of an imaginary unit in a physical equation. Such units arise from one of the many quantities in the real algebra that square to -1, and these have geometric significance because of the properties of the algebra and the interaction of its various subspaces.

In $\mathcal{G}(3,0)$, an exceptional case occurs. Given a standard basis built from orthonourmal e_i's from V, the set of *all* 2-vectors is generated by

$$\{e_3e_2, e_1e_3, e_2e_1\}.$$

Labelling these i, j and k (momentarily deviating from our uppercase convention), the subspace generated by 0-vectors and 2-vectors is exactly $\{\alpha_0 + i\alpha_1 + j\alpha_2 + k\alpha_3 \mid \alpha_i \in \mathbb{R}\}$. This set is seen to be a subalgebra, and furthermore is R-algebra isomorphic to the quaternions, another important algebraic system.

Extensions of The Inner and Outer Products

It is common practice to extend the outer product on vectors to the entire algebra. This may be done through the use of the grade projection operator:

$$C \wedge D := \sum_{r,s} \langle \langle C \rangle_r \langle D \rangle_s \rangle_{r+s} \quad \text{(the \textit{outer product})}$$

The inner product on vectors can also be generalised, but in more than one non-equivalent way. The paper (Dorst 2002) gives a full treatment of several different inner products developed for geometric algebras and their interrelationships. Among these several different generalisations are:

$$C \rfloor D := \sum_{r,s} \langle \langle C \rangle_r \langle D \rangle_s \rangle_{s-r} \quad \text{(the \textit{left contraction})}$$

$$C \lfloor D := \sum_{r,s} \langle \langle C \rangle_r \langle D \rangle_s \rangle_{r-s} \quad \text{(the \textit{right contraction})}$$

$$C * D := \sum_{r,s} \langle \langle C \rangle_r \langle D \rangle_s \rangle_0 \quad \text{(the \textit{scalar product})}$$

$$C \bullet D := \sum_{r,s} \langle \langle C \rangle_r \langle D \rangle_s \rangle_{|s-r|} \text{ (the "(fat) dot" product).}$$

(Dorst 2002) makes an argument for the use of contractions in preference to Hestenes's inner product (below); they are algebraically more regular and have cleaner geometric interpretations. A number of identities relating these products are valid without restriction of their inputs. Benefits of using the left contraction as an extension of the inner product on vectors are that the identity $ab = a \cdot b + a \wedge b$ is extended to $aB = a \rfloor B + a \wedge B$ for a vector *a* and any multivector *B*, and that the projection operation $P_b(a) = (a \cdot b^{-1})b$ is extended to $P_B(A) = (A \rfloor B^{-1}) \rfloor B$ for any blades *A* and *B*.

The inner product generalisation introduced by D. Hestenes ignores the scalar component of both inputs, and must be used care with respect to the grades in each input:

$$C \cdot D := \sum_{r \neq 0, s \neq 0} \langle \langle C \rangle_r \langle D \rangle_s \rangle_{|s-r|} \text{ (Hestenes's inner product)}$$

Relationship with Other Formalisms

$\mathcal{G}(3,0)$ may be directly compared to vector algebra.

The even subalgebra of $\mathcal{G}(2,0)$ is isomorphic to the complex numbers, as may be seen by writing a vector P in terms of its components in an orthonourmal basis and left multiplying by the basis vector e_1, yielding

$$Z = e_1 P = e_1(xe_1 + ye_2) = x(1) + y(e_1 e_2)$$

where we identify $i \to e_1 e_2$ since

$$(e_1 e_2)^2 = e_1 e_2 e_1 e_2 = -e_1 e_1 e_2 e_2 = -1$$

Similarly, the even subalgebra of $\mathcal{G}(3,0)$ with basis $\{1, e_2 e_3, e_3 e_1, e_1 e_2\}$ is isomorphic to theq uaternions as may be seen by identifing $i \to -e_2 e_3$, $j \to -e_3 e_1$ and $k \to -e_1 e_2$.

Every associative algebra has a matrix representation; the Pauli matrices are a representation of $\mathcal{G}(3,0)$ and the Dirac matrices are a representation of $\mathcal{G}(1,3)$, showing the equivalence with matrix representations used by physicists.

Rotations

If we have a product of vectors $R = a_1 a_2 \ldots a_r$ *then we denote the reverse as*

$$R^\dagger = (a_1 a_2 \ldots a_r)^\dagger = a_r \ldots a_2 a_1 .$$

As an example, assume that $R = ab$ we get

$$RR^{\dagger} = abba = ab^2a = a^2b^2 = R^{\dagger}R\cdot$$

Scaling R so that $RR^{\dagger} = 1$ then

$$(RvR^{\dagger})^2 = Rv^2R^{\dagger} = v^2RR^{\dagger} = v^2$$

so $RvR^{\dagger}$ leaves the length of v unchanged. We can also show that

$$Rv_1R^{\dagger}\cdot Rv_2R^{\dagger} = v_1\cdot v_2$$

so the transformation $RvR^{\dagger}$ preserves both length and angle. It therefore can be identified as a rotation or rotoreflection; R is called a rotor if it is a proper rotation (as it is if it can be expressed as a product of an even number of vectors) and is an instance of what is known in GA as a *versor* (presumably for historical reasons).

There is a general method for rotating a vector involving the formation of a multivector of the form $R = e^{\frac{-B\theta}{2}}$ that produces an anticlockwise rotation θ in the plane defined by a bivector B.

Rotors are a generalisation of quaternions to n-D spaces.

Geometric Calculus

Geometric Calculus extends the formalism to include differentiation and integration including differential geometry and differential forms.

Essentially, the vector derivative is defined so that the GA version of Green's theorem is true,

$$\oint_A dA\nabla f = \oint_{dA} dxf$$

and then one can write

$$\nabla f = \nabla\cdot f + \nabla \wedge f$$

as a geometric product, effectively generalising Stokes theorem (including the differential forms version of it).

In $1D$ when A is a curve with endpoints a and b, then

$$\oint_A dA\nabla f = \oint_{dA} dxf$$

reduces to

$$\int_a^b dx\nabla f = \int_a^b dx\cdot\nabla f = \int_a^b df = f(b) - f(a)$$

or the fundamental theorem of integral calculus.

Also developed are the concept of vector manifold and geometric integration theory (which generalises Cartan's differential forms).

Conformal Geometric Algebra (CGA)

A compact description of the current state of the art is provided by Bayro-Corrochano and Scheuermann (2010), which also includes further references, in particular to Dorst et al *(2007).Another useful reference is Li (2008).*

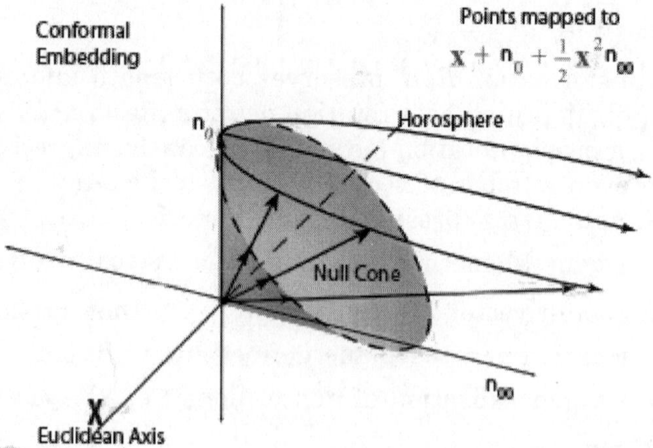

This again extends the language of GA, the conformal model of ε^2 is embedded in the CGA $\mathcal{G}^{4,1} = \mathcal{G}(\mathcal{R}^{4,1})$ via the identification of Euclidean points with $5D$ vectors in the $4D$ null cone, adding a point at infinity and normalising all points to the hyperplane $X{\cdot}n_\infty = -1$. Allows all of conformal algebra to be done by combinations of rotations and reflections and the language is covariant, permitting the extension of incidence relations of projective geometry to circles and spheres.

Specifically, we add e_+ such that $e_+^2 = 1$ and e_- such that $e_-^2 = -1$ to the orthonourmal basis of $\mathcal{G}(3,0)$ allowing the creation of $n_\infty = e_+ + e_-$ representing an ideal point (point at infinity) $n_0 = \frac{1}{2}(e_+ - e_-)$ representing the origin where,

$n_0{\cdot}n_\infty = -1$ and $n_\infty \wedge n_0 = e_+ \wedge e_- = e_+e_- = E$ where $E \in \mathcal{R}(1,1)$. is a unit pseudoscalar representing the Minkowski plane.

This procedure has some similarities to the procedure for working with homogeneous coordinates in projective geometry and in this case allows the modelling of Euclidean transformations as Orthogonal transformations.

A fast changing and fluid area of GA, CGA is also being investigated for applications to relativistic physics.

Software

GA is a very application oriented subject. There is a reasonably steep initial learning curve associated with it, but this can be eased somewhat by the use of applicable software. The following is a list of freely available software that does not require ownership of commercial software or purchase of any commercial products for this purpose:

GA Viewer Fontijne, Dorst, Bouma & Mann : The link provides a manual, introduction to GA and sample material as well as the software.

CLUViz Perwass : Software allowing script creation and including sample visualisations, manual and GA introduction.

Gaigen Fontijne : For programmers, this is a code generator with support for C,C++,C# and Java.

Cinderella Visualisations Hitzer and Dorst.

Algebraic Examples

Area of Parallelogram Spanned by Two Vectors

If A_r is a r-blade then a vector a has a projection or parallel component $P^{||}$ onto A_r,

$$P^{||} = (a \cdot A_r)A_r^{-1}$$

and a rejection or perpendicular component

$$P^{\perp} = (a \wedge A_r)A_r^{-1}$$

So for vectors a and b in 2D we have

$$(a \wedge b)b^{-1} = a^{\perp} \text{ or } a \wedge b = a^{\perp}b$$

and we have that a Ω b is the product of the "altitude" and the "base" of the ab-parallelogram, that is, its area.

Intersection of a Line and a Plane

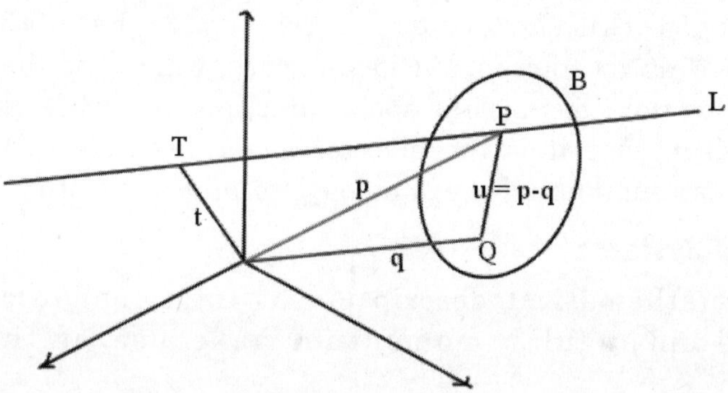

Consider a line L defined by points T and P (which we seek) and a plane defined by a bivector B containing points P and Q.

We may define the line parametrically by $a \wedge b = a^{\perp}b$ where p and t are position vectors for points T and P and v is the direction vector for the line.

Then

$$B \wedge (p - q) = 0 \text{ and } B \wedge (t + \alpha v - q) = 0$$

so

$$\alpha = \frac{B \wedge (q - t)}{B \wedge v}$$

and

$$p = t + \left(\frac{B \wedge (q - t)}{B \wedge v}\right)v.$$

Electrodynamics and Special Relativity

In physics, the main applications are the geometric algebra of Euclidean 3-space, Cl_3, called the Algebra of physical space (APS), and the geometric algebra of Minkowski 3+1 spacetime, $Cl_{3,1}$, called spacetime algebra (STA).

In APS, points of (3+1)-dimensional space-time are represented by paravectors: a 3-dimensional vector (space) plus a 1-dimensional scalar (time), while in STA points of space-time are represented simply by vectors.

In spacetime algebra the electromagnetic field tensor has a bivector representation $F = E + icB$ where the imaginary unit is the volume element, and where Maxwell's equations simplify to one equation:

$$\nabla F = \mu_0 cJ$$

Boosts in this Lorenzian metric space have the same expression e^{β} as rotation in Euclidean space, where β is the bivector generated by the time and the space directions involved, whereas in the Euclidean case it is the bivector generated by the two space directions, strengthening the "analogy" to almost identity.

Rotational Systems

The mathematical description of rotational forces such as Torque and angular momentum make use of the Cross product.

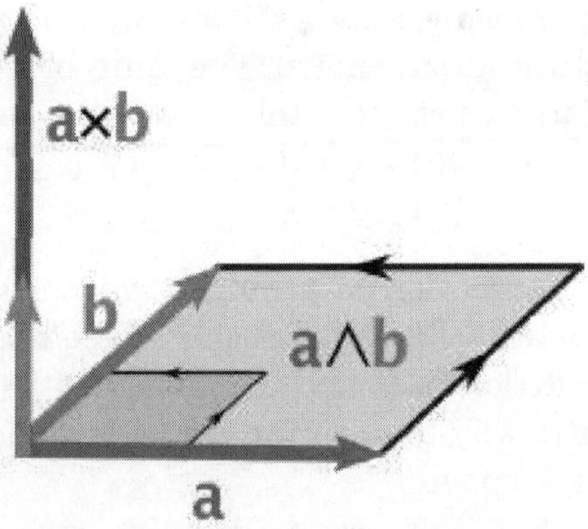

Figure 1: The cross product in relation to the outer product. In red are the unit normal vector, and the "parallel" unit bivector.

The cross product can be viewed in terms of the outer product allowing a more natural geometric interpretation of the cross product as a bivector using the dual relationship

$$a \times b = I(a \wedge b).$$

For example, torque is generally defined as the magnitude of the perpendicular force component times distance, or work per unit angle.

Suppose a circular path in an arbitrary plane containing orthonourmal vectors $\hat{u}$ and $\hat{v}$ is parameterised by angle.

$$r = r(\hat{u}\cos\theta + \hat{v}\sin\theta) = r\hat{u}(\cos\theta + \hat{u}\hat{v}\sin\theta)$$

By designating the unit bivector of this plane as the imaginary number

$$i = \hat{u}\hat{v} = \hat{u} \wedge \hat{v}$$
$$i^2 = -1$$

this path vector can be conveniently written in complex exponential form

$$r = r\hat{u}e^{i\theta}$$

and the derivative with respect to angle is

$$\frac{dr}{d\theta} = r\hat{u}ie^{i\theta} = ri$$

So the torque, the rate of change of work *W*, due to a force *F*, is

$$\tau = \frac{dW}{d\theta} = F\cdot\frac{dr}{d\theta} = F\cdot(ri)$$

Unlike the cross product description of torque, $\tau = r \times F$, the geometric-algebra description does not introduce a vector in the normal direction; a vector that does not exist in two and that is not unique in greater than three dimensions. The unit bivector describes the plane and the orientation of the rotation, and the sense of the rotation is relative to the angle between the vectors $\hat{u}$ and $\hat{v}$.

History

Although the connection of geometry with algebra dates as far back at least to Euclid's Elements in the 3rd century B.C., GA in the sense used in this article was not developed until 1844, when it was used in a systematic way to describe the geometrical properties and transformations of a space.

In that year, Hermann Grassmann introduced the idea of a geometrical algebra in full generality as a certain calculus (analogous to the propositional calculus) which encoded all of the geometrical information of a space. Grassmann's algebraic system could be applied to a number of different kinds of spaces: the chief among them being Euclidean space, affine space, and projective space. Following Grassmann, in 1878 William Kingdon Clifford examined Grassmann's algebraic system alongside the quaternions of William Rowan Hamilton in (Clifford 1878).

From his point of view, the quaternions described certain transformations (which he called rotors), whereas Grassmann's algebra described certain properties (or Strecken such as length, area, and volume). His contribution was to define a new product — the geometric product — on an existing Grassmann algebra, which realised the quaternions as living within that algebra. Subsequently Rudolf Lipschitz in 1886 generalised Clifford's interpretation of the quaternions and applied them to the geometry of rotations in n dimensions. Later these developments would lead other 20th-century mathematicians to formalise and explore the properties of the Clifford algebra.

Nevertheless, another revolutionary development of the 19th-century would completely overshadow the geometric algebras: that of vector analysis, developed independently by Josiah Willard Gibbs andOliver Heaviside. Vector analysis was motivated by James Clerk Maxwell's studies of electromagnetism, and specifically the need to express and manipulate conveniently certain differential equations. Vector analysis had a certain intuitive appeal compared to the rigours of the new algebras. Physicists and mathematicians alike readily adopted it as their geometrical toolkit of choice, particularly following the influential 1901 textbook Vector Analysis by Edwin Bidwell Wilson, following lectures of Gibbs.

In more detail, there have been three approaches to geometric algebra: quaternionic analysis, initiated by Hamilton in 1843 and geometrised as rotors by Clifford in 1878; geometric algebra, initiated by Grassmann in 1844; and vector analysis, developed out of quaternionic analysis in the late 19th century by Gibbs and Heaviside.

The legacy of quaternionic analysis in vector analysis can be seen in the use of i,j,k to indicate the basis vectors of R^3: it is being thought of as the purely imaginary quaternions.

From the perspective of geometric algebra, quaternions can be identified as $C-\ell_{3,0}(R)$, the even part of the Clifford algebra on Euclidean 3-space, which unifies the three approaches.

20th Century and Present

Progress on the study of Clifford algebras quietly advanced through the twentieth century, although largely due to the work of abstract algebraists such as Hermann Weyl and Claude Chevalley. The *geometrical* approach to geometric algebras has seen a number of 20th-century revivals.

In mathematics, Emil Artin's *Geometric Algebra* discusses the algebra associated with each of a number of geometries, including affine geometry, projective geometry, symplectic geometry, and orthogonal geometry.

In physics, geometric algebras have been revived as a "new" way to do classical mechanics and electromagnetism, together with more advanced topics such as quantum mechanics and gauge theory. David Hestenes reinterpreted the Pauli and Dirac matrices as vectors in ordinary space and spacetime, respectively, and has been a primary contemporary advocate for the use of geometric algebra.

In computer graphics, geometric algebras have been revived in order to represent efficiently rotations (and other transformations) on computer hardware.

Max-plus Algebra

A max-plus algebra is an algebra over the real numbers with maximum and addition as the two binary operations. It can be used appropriately to determine marking times within a given Petri net and a vector filled with marking state at the beginning.

Operators

Scalar Operations: Let *A* and *B* be scalars. Then the operations maximum (implied by the max operator $\oplus$) and addition (plus operator $\oplus$) for this scalars are defined as

$$A \oplus B = \max(A, B)$$

$$A \otimes B = A + B$$

Watch: Max-operator $\oplus$ can easily be confused with the addition operation. All $\oplus$ - operations have a higher precedence than $\oplus$ - operations.

Matrix Operations

Max-Plus algebra can be used for matrix operands *M*, *N* likewise. To perform the $R = M \oplus N$ - operation, the elements of the resulting matrix R (row i, column j) have to be set up by the maximum operation of both corresponding elements of the matrices *M* and *N*:

$$R_{ij} = M_{ij} \oplus N_{ij}$$

The $\oplus$ - operation is similar to algorithm of Matrix multiplication, however, every "+" calculation has to be substituted by a $\oplus$ - operation, every "." calculation by a - operation.

Useful enhancement elements

In order to handle marking times like $-\infty$ which means "never before", the ε-element has been established by $\varepsilon = -\infty$. According to the idea of infinity, the following equations can be found:

$$\varepsilon \oplus A = A$$

$$\varepsilon \ A = \varepsilon$$

To point the zero number out, the element *e* was defined by $e = 0$. Therefore:

$$e \oplus A = A$$

Obviously, ε is the neutral element for the $\oplus$ - operation as well as *e* is for the $\oplus$ - operation

Algebra Properties

- associativity:

$$(A \oplus B) \oplus C = A \oplus (B \oplus C)$$

$$(A \otimes B) \otimes C = A \otimes (B \otimes C)$$

- commutativity :

$$A \oplus B = B \oplus A$$

- distributivity:

$$(A \oplus B) \otimes C = A \otimes C \oplus B \otimes C$$

Note: In general, A $\oplus$ B = B $\oplus$ A does not hold, for example in the case of matrix operations.

Associative Algebra

In mathematics, an associative algebra *A* is an associative ring that has a compatible structure of a vector space over a certain field *K* or, more generally, of a module over a commutative ring *R*. Thus *A*is endowed with binary operations of addition and multiplication satisfying a number of axioms, including associativity of multiplication and distributivity, as well as compatible multiplication by the elements of the field *K* or the ring *R*.

In some areas of mathematics, associative algebras are typically assumed to have a multiplicative unit, denoted 1. To make this extra assumption clear, these associative algebras are called unital algebras.

Formal Definition

Let R *be a fixed commutative ring. An associative* R*-algebra is an additive abelian group* A *which has the structure of both a ring and an* R*-module in such a way that ring multiplication is* R*-bilinear:*

$$r \cdot (xy) = (r \cdot x)y = x(r \cdot y)$$

for all $r \in R$ and $x, y \in A$. We say *A* is unital if it contains an element 1 such that

$$1x = x = x1$$

for all $x \in A$.

If *A* itself is commutative (as a ring) then it is called a commutative *R*-algebra.

From R-modules

Starting with an *R*-module *A*, we get an associative *R*-algebra by equipping *A* with an *R*-bilinear mapping $A \times A \to A$ such that

$$x(yz) = (xy)z$$

for all *x*, *y*, and *z* in *A*. This *R*-bilinear mapping then gives *A* the structure of a ring and an associative R-algebra. Every associative *R*-algebra arises this way.

Moreover, the algebra A built this way will be unital if and only if

$$\exists 1 \in A, 1x = x1 = x$$

This definition is equivalent to the statement that a unital associative R-algebra is a monoid in R-Mod (the monoidal category of R-modules).

From Rings

Starting with a ring A, we get a unital associative R-algebra by providing a ring homomorphism $\eta : R \to A$ whose image lies in the centre of A. The algebra A can then be thought of as an R-module by defining

$$r \cdot x = \eta(r)x$$

for all $r \in R$ and $x \in A$.

If A is commutative then the centre of A is equal to A, so that a commutative R-algebra can be defined simply as a homomorphism $\eta : R \to A$ of commutative rings.

Algebra Homomorphisms

A homomorphism between two associative R-*algebras is an* R-*linear ring homomorphism. Explicitly,* $\phi : A_1 \to A_2$ *is an associative algebra homomorphism if*

$$\phi(r \cdot x) = r \cdot \phi(x)$$

$$\phi(x + y) = \phi(x) + \phi(y)$$

$$\phi(xy) = \phi(x)\phi(y)$$

For a homomorphism of *unital* associative R-algebras, we also demand that

$$\phi(1) = 1$$

The class of all unital associative R-algebras together with algebra homomorphisms between them form a category, sometimes denoted R-Alg.

The subcategory of commutative R-algebras can be characterised as the coslice category R/C Ring where C Ring is the category of commutative rings.

Examples

- The square n-by-n matrices with entries from the field K form a unitary associative algebra over K.

- The complex numbers form a 2-dimensional unitary associative algebra over the real numbers.
- The quaternions form a 4-dimensional unitary associative algebra over the reals (but not an algebra over the complex numbers, since if complex numbers are treated as a subset of the quaternions, complex numbers and quaternions do not commute).
- The 2 × 2 real matrices form an associative algebra useful in plane mapping.
- The polynomials with real coefficients form a unitary associative algebra over the reals.
- Given any Banach space X, the continuous linear operators $A : X \to X$ form a unitary associative algebra (using composition of operators as multiplication); this is a Banach algebra.
- Given any topological space X, the continuous real- or complex-valued functions on X form a real or complex unitary associative algebra; here the functions are added and multiplied pointwise.
- An example of a non-unitary associative algebra is given by the set of all functions f: R → R whose limit as x nears infinity is zero.
- The Clifford algebras, which are useful in geometry and physics.
- Incidence algebras of locally finite partially ordered sets are unitary associative algebras considered in combinatorics.
- Any ring A can be considered as a Z-algebra in a unique way. The unique ring homomorphism from Z to A is determined by the fact that it must send 1 to the identity in A. Therefore rings and Z-algebras are equivalent concepts, in the same way that abelian groups and Z-modules are equivalent.
- Any ring of characteristic n is a (Z/nZ)-algebra in the same way.
- Any ring A is an algebra over its centre $Z(A)$, or over any subring of its centre.
- Any commutative ring R is an algebra over itself, or any subring of R.
- Given an R-module M, the endomorphism ring of M, denoted $\text{End}_R(M)$ is an R-algebra by defining $(r \cdot \varphi)(x) = r \cdot \varphi(x)$.

- Any ring of matrices with coefficients in a commutative ring R forms an R-algebra under matrix addition and multiplication. This coincides with the previous example when M is a finitely-generated, free R-module.
- Every polynomial ring $R[x_1, ..., x_n]$ is a commutative R-algebra. In fact, this is the free commutative R-algebra on the set $\{x_1, ..., x_n\}$.
- The free R-algebra on a set E is an algebra of polynomials with coefficients in R and noncommuting indeterminates taken from the set E.
- The tensor algebra of an R-module is naturally an R-algebra. The same is true for quotients such as the exterior and symmetric algebras. Categorically speaking, the functor which maps an R-module to its tensor algebra is left adjoint to the functor which sends an R-algebra to its underlying R-module (forgetting the ring structure).
- Given a commutative ring R and any ring A the tensor product $R\Delta_Z A$ can be given the structure of an R-algebra by defining $r \cdot (s\Delta a) = (rs\Delta a)$. The functor which sends A to $R\Delta_Z A$ is left adjoint to the functor which sends an R-algebra to its underlying ring (forgetting the module structure).

Constructions

Subalgebras: A subalgebra of an R-algebra A is a subset of A which is both a subring and a submodule of A. That is, it must be closed under addition, ring multiplication, scalar multiplication, and it must contain the identity element of A.

Quotient Algebras: Let A be an R-algebra. Any ring-theoretic ideal I in A is automatically an R-module since $r \cdot x = (r1_A)x$.

This gives the quotient ring A/I the structure of an R-module and, in fact, an R-algebra. It follows that any ring homomorphic image of A is also an R-algebra.

Direct Products

The direct product of a family of R-algebras is the ring-theoretic direct product. This becomes an R-algebra with the obvious scalar multiplication.

Free Products

One can form a free product of R-algebras in a manner similar to the free product of groups. The free product is the coproduct in the category of R-algebras.

Tensor Products

The tensor product of two R-algebras is also an R-algebra in a natural way.

Associativity and The Multiplication Mapping

Associativity was defined above quantifying over all elements *of* A. *It is possible to define associativity in a way that does not explicitly refer to elements. An algebra is defined as a map* M *(multiplication) on a vector space* A:

$$M : A \times A \rightarrow A$$

An associative algebra is an algebra where the map M has the property

$$M^{\circ}(Id \times M) = M^{\circ}(M \times Id)$$

Here, the symbol $\circ$ refers to function composition, and Id : $A \rightarrow A$ is the identity map on A.

We need only understand that each side of the above equation is a function that takes three arguments. For example, the left-hand side acts as

$$(M^{\circ}(Id \times M))(x,y,z) = M(x,M(y,z))$$

Similarly, a unital associative algebra can be defined in terms of a unit map

$$\eta : K \rightarrow A$$

which has the property

$$M^{\circ}(Id \times \eta) = s = M^{\circ}(\eta \times Id)$$

Here, the unit map η takes an element k in K to the element $k1$ in A, where 1 is the unit element of A. The map s is just plain-old scalar multiplication: $s : K \times A \rightarrow A$; thus, the above identity is sometimes written with Id standing in the place of s, with scalar multiplication being implicitly understood.

Coalgebras

An associative unitary algebra over K is based on a morphism $A \times A \rightarrow A$ having 2 inputs (multiplicator and multiplicand) and one output (product), as well as a morphism $K \rightarrow A$ identifying the scalar multiples of the multiplicative identity. These two morphisms can be dualized using categorial duality by reversing all arrows in the commutative diagrams which describe the algebra axioms; this defines the structure of a coalgebra.

There is also an abstract notion of F-coalgebra.

Representations

A representation of an algebra is a linear map $\rho: A \to gl(V)$ from A to the general linear algebra of some vector space (or module) V that preserves the multiplicative operation: that is, $\rho(xy)=\rho(x)\rho(y)$.

Note, however, that there is no natural way of defining a tensor product of representations of associative algebras, without somehow imposing additional conditions. Here, by tensor product of representations, the usual meaning is intended: the result should be a linear representation on the product vector space. Imposing such additional structure typically leads to the idea of a Hopf algebraor a Lie algebra, as demonstrated below.

Motivation for a Hopf Algebra

Consider, for example, two representations $\sigma : A \to gl(V)$ and $\tau : A \to gl(W)$. One might try to form a tensor product representation $\rho : x \mapsto \sigma(x) \otimes \tau(x)$ according to how it acts on the product vector space, so that

$$\rho(x)(v \otimes w) = (\sigma(x)(v)) \otimes (\tau(x)(w)).$$

However, such a map would not be linear, since one would have

$$\rho(kx) = \sigma(kx) \otimes \tau(kx) = k\sigma(x) \otimes k\tau(x) = k^2(\sigma(x) \otimes \tau(x)) = k^2\rho(x)$$

for $k \in K$. One can rescue this attempt and restore linearity by imposing additional structure, by defining a map $\Delta: A \to A \times A$, and defining the tensor product representation as

$$\rho = (\sigma \otimes \tau)^\circ \Delta.$$

Here, Δ is a comultiplication. The resulting structure is called a bialgebra. To be consistent with the definitions of the associative algebra, the coalgebra must be co-associative, and, if the algebra is unital, then the co-algebra must be unital as well. Note that bialgebras leave multiplication and co-multiplication unrelated; thus it is common to relate the two (by defining an antipode), thus creating a Hopf algebra.

Motivation for a Lie Algebra

One can try to be more clever in defining a tensor product. Consider, for example,

$$x \mapsto \rho(x) = \sigma(x) \otimes \mathrm{Id}_W + \mathrm{Id}_V \otimes \tau(x)$$

so that the action on the tensor product space is given by

$$\rho(x)(v \otimes w) = (\sigma(x)v) \otimes w + v \otimes (\tau(x)w) .$$

This map is clearly linear in *x*, and so it does not have the problem of the earlier definition. However, it fails to preserve multiplication:

$$\rho(xy) = \sigma(x)\sigma(y) \otimes \mathrm{Id}_W + \mathrm{Id}_V \otimes \tau(x)\tau(y) .$$

But, in general, this does not equal

$$\rho(x)\rho(y) = \sigma(x)\sigma(y) \otimes \mathrm{Id}_W + \sigma(x) \otimes \tau(y) + \sigma(y) \otimes \tau(x) + \mathrm{Id}_V \otimes \tau(x)\tau(y) .$$

Equality would hold if the product *xy* were antisymmetric (if the product were the Lie bracket, that is, $xy \equiv M(x,y) = [x,y]$), thus turning the associative algebra into a Lie algebra.

Lie Algebra

Definition and First Properties

A Lie algebra is a vector space $\mathfrak{g}$ *over some field* F together *with a binary operation [·, ·]*

$$[\cdot,\cdot] : \mathfrak{g} \times \mathfrak{g} \to \mathfrak{g}$$

called the Lie bracket, which satisfies the following axioms:

- Bilinearity:

$$[ax + by, z] = a[x,z] + b[y,z], \quad [z, ax + by] = a[z,x] + b[z,y]$$

for all scalars *a*, *b* in *F* and all elements *x*, *y*, *z* in $\mathfrak{g}$.

- Alternating on $\mathfrak{g}$:

$$[x,x] = 0$$

for all *x* in $\mathfrak{g}$.

- The Jacobi identity:

$$[x,[y,z]] + [y,[z,x]] + [z,[x,y]] = 0$$

for all *x*, *y*, *z* in $\mathfrak{g}$.

Note that the bilinearity and alternating properties imply anticommutativity, i.e., $[x,y] = -[y,x]$ for all elements *x*, *y* in $\mathfrak{g}$, while anticommutativity only implies the alternating property if the field's characteristic is not 2.

For any associative algebra A with multiplication $*$, one can construct a Lie algebra $L(A)$. As a vector space, $L(A)$ is the same as A. The Lie bracket of two elements of $L(A)$ is defined to be their commutator in A:

$$[a,b] = a * b - b * a.$$

The associativity of the multiplication $*$ in A implies the Jacobi identity of the commutator in $L(A)$. In particular, the associative algebra of $n \times n$ matrices over a field F gives rise to the general linear Lie algebra $\mathfrak{gl}_n(F)$. The associative algebra A is called an enveloping algebra of the Lie algebra $L(A)$. It is known that every Lie algebra can be embedded into one that arises from an associative algebra in this fashion.

Homomorphisms, Subalgebras, and Ideals

The Lie bracket is not an associative operation in general, meaning that $[[x,y],z]$ need not equal$[x,[y,z]]$. Nonetheless, much of the terminology that was developed in the theory of associative ringsor associative algebras is commonly applied to Lie algebras. A subspace $\mathfrak{h} \subseteq \mathfrak{g}$ that is closed under the Lie bracket is called a Lie subalgebra. If a subspace $I \subseteq \mathfrak{g}$ satisfies a stronger condition that

$$[\mathfrak{g}, I] \subseteq I,$$

then I is called an ideal in the Lie algebra $\mathfrak{g}$. A Lie algebra in which the commutator is not identically zero and which has no proper ideals is called simple. A homomorphism between two Lie algebras (over the same ground field) is a linear map that is compatible with the commutators:

$$f : \mathfrak{g} \to \mathfrak{g}', \quad f([x,y]) = [f(x), f(y)],$$

for all elements x and y in $\mathfrak{g}$. As in the theory of associative rings, ideals are precisely the kernels of homomorphisms, given a Lie algebra $\mathfrak{g}$ and an ideal I in it, one constructs the factor algebra $\mathfrak{g}/I$, and the first isomorphism theorem holds for Lie algebras. Given two Lie algebras $\mathfrak{g}$ and $\mathfrak{g}'$, their direct sum is the vector space $\mathfrak{g} \oplus \mathfrak{g}'$ consisting of the pairs $(x,x'), x \in \mathfrak{g}, x' \in \mathfrak{g}'$, with the operation

$$[(x,x'),(y,y')] = ([x,y],[x',y']), \quad x,y \in \mathfrak{g}, x',y' \in \mathfrak{g}'.$$

Examples

Any vector space *V* endowed with the identically zero Lie bracket becomes a Lie algebra. Such Lie algebras are called abelian, cf. below. Any one-dimensional Lie algebra over a field is abelian, by the antisymmetry of the Lie bracket.

The three-dimensional Euclidean space R^3 with the Lie bracket given by the cross product off vecors becomes a three-dimensional Lie algebra.

The Heisenberg algebra is a three-dimensional Lie algebra with generators:

$$x=\begin{pmatrix}0&1&0\\0&0&0\\0&0&0\end{pmatrix},\quad y=\begin{pmatrix}0&0&0\\0&0&1\\0&0&0\end{pmatrix},\quad z=\begin{pmatrix}0&0&1\\0&0&0\\0&0&0\end{pmatrix},$$

whose commutation relations are

$$[x,y]=z,\quad [x,z]=0,\quad [y,z]=0.$$

It is explicitly exhibited as the space of 3×3 strictly upper-triangular matrices.

The subspace of the general linear Lie algebra $\mathfrak{gl}_n(F)$ consisting of matrices of trace zero is a subalgebra, the special linear Lie algebra, denoted $\mathfrak{sl}_n(F)$.

Any Lie group *G* defines an associated real Lie algebra $\mathfrak{g}=\text{Lie}(G)$. The definition in general is somewhat technical, but in the case of real matrix groups, it can be formulated via the exponential map, or the matrix exponent. The Lie algebra $\mathfrak{g}$ consists of those matrices *X* for which

$$\exp(tX)\in G$$

for all real numbers *t*. The Lie bracket of $\mathfrak{g}$ is given by the commutator of matrices. As a concrete example, consider the special linear group SL(*n*,R), consisting of all $n \times n$ matrices with real entries and determinant 1. This is a matrix Lie group, and its Lie algebra consists of all $n \times n$ *matrices* with real entries and trace 0.

The real vector space of all $n \times n$ skew-hermitian matrices is closed under the commutator and forms a real Lie algebra denoted $\mathfrak{u}(n)$. This is the Lie algebra of the unitary group *U*(*n*).

An important class of infinite-dimensional real Lie algebras arises in differential topology. The space of smooth vector fields on a differentiable manifold M forms a Lie algebra, where the Lie bracket is defined to be the commutator of vector fields. One way of expressing the Lie bracket is through the formalism of Lie derivatives, which identifies a vector field X with a first order partial differential operator L_X acting on smooth functions by letting $L_X(f)$ be the directional derivative of the function f in the direction of X. The Lie bracket $[X, Y]$ of two vector fields is the vector field defined through its action on functions by the formula:

$$L_{[X,Y]}f = L_X(L_Yf) - L_Y(L_Xf).$$

This Lie algebra is related to the pseudogroup of diffeomorphisms of M.

The commutation relations between the x, y, and z components of the angular momentum operator in quantum mechanics form a representation of a complex three-dimensional Lie algebra, which is the complexification of the Lie algebra $so(3)$ of the three-dimensional rotation group:

$$[L_x, L_y] = i\hbar L_z$$

$$[L_y, L_z] = i\hbar L_x$$

$$[L_z, L_x] = i\hbar L_y$$

Kac–Moody algebra is an example of an infinite-dimensional Lie algebra.

Structure Theory and Classification

Every finite-dimensional real or complex Lie algebra has a faithful representation by matrices (Ado's theorem). Lie's fundamental theorems describe a relation between Lie groups and Lie algebras. In particular, any Lie group gives rise to a canonically determined Lie algebra (concretely, the tangent space at the identity), and conversely, for any Lie algebra there is a corresponding connected Lie group (Lie's third theorem). This Lie group is not determined uniquely, however, any two connected Lie groups with the same Lie algebra are locally isomorphic, and in particular, have the same universal cover. For instance, the special orthogonal group SO(3) and the special unitary group SU(2) give rise to the same Lie algebra, which is isomorphic to R^3 with the cross-product, and SU(2) is a simply-connected twofold cover of SO(3). Real and complex Lie algebras can be classified to some extent, and this is often an important step toward the classification of Lie groups.

Abelian, Nilpotent, and Solvable

Analogously to abelian, nilpotent, and solvable groups, defined in terms of the derived subgroups, one can define abelian, nilpotent, and solvable Lie algebras.

A Lie algebra $\mathfrak{g}$ is abelian if the Lie bracket vanishes, i.e. $[x,y] = 0$, for all x and y in $\mathfrak{g}$. Abelian Lie algebras correspond to commutative (or abelian) connected Lie groups such as vector spaces K^n ortori T^n, and are all of the form $\mathfrak{k}^n$, meaning an n-dimensional vector space with the trivial Lie bracket.

A more general class of Lie algebras is defined by the vanishing of all commutators of given length. A Lie algebra $\mathfrak{g}$ is nilpotent if the lower central series

$$\mathfrak{g} > [\mathfrak{g},\mathfrak{g}] > [[\mathfrak{g},\mathfrak{g}],\mathfrak{g}] > [[[\mathfrak{g},\mathfrak{g}],\mathfrak{g}],\mathfrak{g}] > \cdots$$

becomes zero eventually. By Engel's theorem, a Lie algebra is nilpotent if and only if for every u in $\mathfrak{g}$ the adjoint endomorphism

$$\mathrm{ad(u)} : \mathfrak{g} \to \mathfrak{g}, \quad \mathrm{ad(u)v} = [\mathrm{u},\mathrm{v}]$$

is nilpotent.

More generally still, a Lie algebra $\mathfrak{g}$ is said to be solvable if the derived series:

$$\mathfrak{g} > [\mathfrak{g},\mathfrak{g}] > [[\mathfrak{g},\mathfrak{g}],[\mathfrak{g},\mathfrak{g}]] > [[[\mathfrak{g},\mathfrak{g}],[\mathfrak{g},\mathfrak{g}]],[[\mathfrak{g},\mathfrak{g}],[\mathfrak{g},\mathfrak{g}]]] > \cdots$$

becomes zero eventually.

Every finite-dimensional Lie algebra has a unique maximal solvable ideal, called its radical. Under the Lie correspondence, nilpotent (respectively, solvable) connected Lie groups correspond to nilpotent (respectively, solvable) Lie algebras.

Simple and Semisimple

A Lie algebra is "simple" if it has no non-trivial ideals and is not abelian. A Lie algebra $\mathfrak{g}$ is called semisimple if its radical is zero. Equivalently, $\mathfrak{g}$ is semisimple if it does not contain any non-zero abelian ideals. In particular, a simple Lie algebra is semisimple. Conversely, it can be proven that any semisimple Lie algebra is the direct sum of its minimal ideals, which are canonically determined simple Lie algebras.

The concept of semisimplicity for Lie algebras is closely related with the complete reducibility of their representations. When the ground field F has characteristic zero, semisimplicity of a Lie algebra $\mathfrak{g}$ over F is equivalent to the complete reducibility of all finite-dimensional representations of $\mathfrak{g}$ An early proof of this statement proceeded via connection with compact groups (Weyl's unitary trick), but later entirely algebraic proofs were found.

Classification

In many ways, the classes of semisimple and solvable Lie algebras are at the opposite ends of the full spectrum of the Lie algebras. The Levi decomposition expresses an arbitrary Lie algebra as a semidirect sum of its solvable radical and a semisimple Lie algebra, almost in a canonical way. Semisimple Lie algebras over an algebraically closed field have been completely classified through their root systems. The classification of solvable Lie algebras is a 'wild' problem, and cannot be accomplished in general.

Cartan's criterion gives conditions for a Lie algebra to be nilpotent, solvable, or semisimple. It is based on the notion of the Killing form, a symmetric bilinear form on $\mathfrak{g}$ defined by the formula

$$K(u,v) = tr(ad(u)ad(v)),$$

where tr denotes the trace of a linear operator. A Lie algebra $\mathfrak{g}$ is semisimple if and only if the Killing form is nondegenerate. A Lie algebra $\mathfrak{g}$ is solvable if and only if $K(\mathfrak{g},[\mathfrak{g},\mathfrak{g}]) = 0$.

Relation to Lie Groups

Although Lie algebras are often studied in their own right, historically they arose as a means to study Lie groups. Given a Lie group, a Lie algebra can be associated to it either by endowing the tangent space to the identity with the differential of the adjoint map, or by considering the left-invariant vector fields as mentioned in the examples. This association is functorial, meaning that homomorphisms of Lie groups lift to homomorphisms of Lie algebras, and various properties are satisfied by this lifting: it commutes with composition, it maps Lie subgroups, kernels, quotients and cokernels of Lie groups to subalgebras, kernels, quotients and cokernels of Lie algebras, respectively.

The functor which takes each Lie group to its Lie algebra and each homomorphism to its differential is a faithful and exact functor. This functor is not invertible; different Lie groups may have the same Lie algebra, for example SO(3) and SU(2) have isomorphic Lie algebras. Even worse, some Lie algebras need not have *any* associated Lie group. Nevertheless, when the Lie algebra is finite-dimensional, there is always at least one Lie group whose Lie algebra is the one under discussion, and a preferred Lie group can be chosen. Any finite-dimensional connected Lie group has a universal cover. This group can be constructed as the image of the Lie algebra under the exponential map. More generally, we have that the Lie algebra is homeomorphic to a neighbourhood of the identity. But globally, if the Lie group is compact, the exponential will not be injective, and if the Lie group is not connected, simply connected or compact, the exponential map need not be surjective.

If the Lie algebra is infinite-dimensional, the issue is more subtle. In many instances, the exponential map is not even locally a homeomorphism (for example, in Diff(S^1), one may find diffeomorphisms arbitrarily close to the identity which are not in the image of exp). Furthermore, some infinite-dimensional Lie algebras are not the Lie algebra of any group.

The correspondence between Lie algebras and Lie groups is used in several ways, including in the classification of Lie groups and the related matter of the representation theory of Lie groups. Every representation of a Lie algebra lifts uniquely to a representation of the corresponding connected, simply connected Lie group, and conversely every representation of any Lie group induces a representation of the group's Lie algebra; the representations are in one to one correspondence. Therefore, knowing the representations of a Lie algebra settles the question of representations of the group. As for classification, it can be shown that any connected Lie group with a given Lie algebra is isomorphic to the universal cover mod a discrete central subgroup. So classifying Lie groups becomes simply a matter of counting the discrete subgroups of the centre, once the classification of Lie algebras is known (solved by Cartan et al. in the semisimple case).

Category Theoretic Definition

Using the language of category theory, a Lie algebra can be defined as an object A *in* Vec_k, *the category of vector spaces over a field* k *of characteristic not 2, together with a morphism* [.,.]: A Δ A→ A, *where* ⊗ *refers to the monoidal product of* Vec_k, *such that*

- $[\cdot,\cdot]^\circ(\mathrm{id}+\tau_{A,A})=0$
- $[\cdot,\cdot]^\circ([\cdot,\cdot]\otimes \mathrm{id})^\circ(\mathrm{id}+\sigma+\sigma^2)=0$

where τ (*a* Δ *b*) := *b* Δ *a* and σ is the cyclic permutation braiding (id Δ $\tau_{A,A}$) ° ($\tau_{A,A}$ Δ id). Indiagrammatic form:

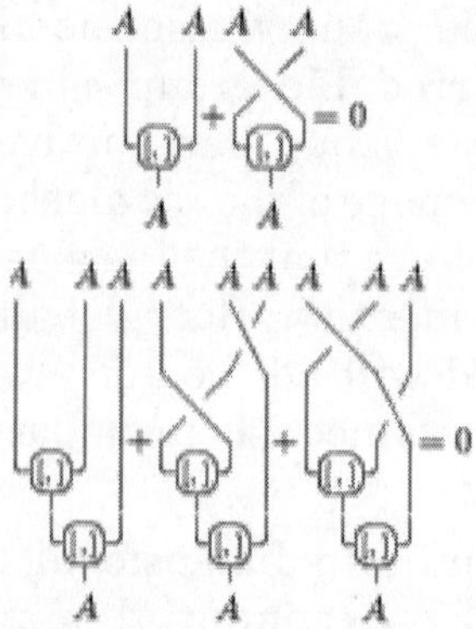

Boolean Algebra (Structure)

In abstract algebra, a Boolean algebra or Boolean lattice is a complemented distributive lattice. This type of algebraic structure captures essential properties of both set operations and logicoperations. A Boolean algebra can be seen as a generalisation of a power set algebra or a field of sets. It is also a special case of De Morgan algebra and Kleene algebra.

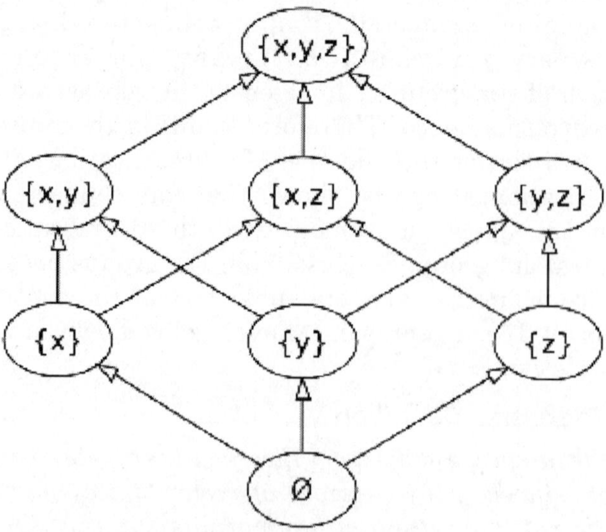

Figure 2: Boolean lattice of subsets

History

The term "Boolean algebra" honours George Boole(1815–1864), a self-educated English mathematician. He introduced the algebraic system initially in a small pamphlet, The Mathematical Analysis of Logic, published in 1847 in response to an ongoing public controversy between Augustus De Morgan and William Hamilton, and later as a more substantial book, The Laws of Thought, published in 1854. Boole's formulation differs from that described above in some important respects. For example, conjunction and disjunction in Boole were not a dual pair of operations. Boolean algebra emerged in the 1860s, in papers written by William Jevons and Charles Sanders Peirce. The first systematic presentation of Boolean algebra and distributive lattices is owed to the 1890 Vorlesungen of Ernst Schrφder. The first extensive treatment of Boolean algebra in English is A. N. Whitehead's 1898 Universal Algebra. Boolean algebra as an axiomatic algebraic structure in the modern axiomatic sense begins with a 1904 paper by Edward Vermilye Huntington. Boolean algebra came of age as serious mathematics with the work of Marshall Stone in the 1930s, and with Garrett Birkhoff's 1940 Lattice Theory. In the 1960s, Paul Cohen, Dana Scott, and others found deep new results in mathematical logic and axiomatic set theory using offshoots of Boolean algebra, namely forcing and Boolean-valued models.

Definition

A Boolean algebra is a six-tuple consisting of a set A, equipped with two binary operations Ω (called "meet" or "and"), (" (called "join" or "or"), a unary operation ¬ (called "complement" or "not") and two elements 0 and 1 (sometimes denoted by the symbols ¥" and £", respectively), such that for all elements a, b and c of A, the following axioms hold:

$a \vee (b \vee c) = (a \vee b) \vee c$	$a \,\Omega\, (b \,\Omega\, c) = (a \,\Omega\, b) \,\Omega\, c$	associativity
$a \vee b = b \vee a$	$a \,\Omega\, b = b \,\Omega\, a$	commutativity
$a \vee (a \vee b) = a$	$a \,\Omega\, (a \vee b) = a$	absorption
$a \vee (b \,\Omega\, c) = (a \vee b) \,\Omega\, (a \vee c)$	$a \,\Omega\, (b \vee c) = (a \,\Omega\, b) \vee (a \,\Omega\, c)$	distributivity
$a \vee \neg a = 1$	$a \,\Omega\, \neg a = 0$	complements

A Boolean algebra with only one element is called a trivial Boolean algebra or a degenerate Boolean algebra. (Some authors require 0 and 1 to be *distinct* elements in order to exclude this case.)

It follows from the first three pairs of axioms above (associativity, commutativity and absorption) that

$$a = a \,\Omega\, b \quad \text{if and only if} \quad a \vee b = b.$$

The relation ≠ defined by $a \neq b$ if and only if the above equivalent conditions hold, is a partial order with least element 0 and greatest element 1. The meet $a \,\Omega\, b$ and the join $a \vee b$ of two elements coincide with their infimum and supremum, respectively, with respect to ≠.

As in every lattice, the relations Ω and ∨ satisfy the first three pairs of axioms above; the fourth pair is just distributivity. Since the complements in a distributive lattice are unique, to define the involution ¬ it suffices to define $\neg a$ as the complement of a.

The set of axioms is self-dual in the sense that if one exchanges ∨ with Ω and 0 with 1 in an axiom, the result is again an axiom. Therefore by applying this operation to a Boolean algebra (or Boolean lattice), one obtains another Boolean algebra with the same elements; it is called its dual.

Homomorphisms and Isomorphisms

A homomorphism *between two Boolean algebras* A *and* B *is a function* f : A → B *such that for all* a, bin A:

$$f(a \vee b) = f(a) \vee f(b),$$

$$f(a\ \Omega\ b) = f(a)\ \Omega\ f(b),$$

$$f(0) = 0,$$

$$f(1) = 1.$$

It then follows that $f(\neg a) = \neg f(a)$ for all a in A as well. The class of all Boolean algebras, together with this notion of morphism, forms a full subcategory of the category of lattices.

Boolean Rings, Ideals and Filters

Every Boolean algebra (A, Ω, ∨) gives rise to a ring (A, +, ·) *by defining* a + b := (a Ω ¬b) ∨ (b Ω ¬a) = (a ∨ b) Ω ¬(a Ω b) *(this operation is called symmetric difference in the case of sets and XOR in the case of logic) and* a · b := a Ω b. *The zero element of this ring coincides with the 0 of the Boolean algebra; the multiplicative identity element of the ring is the 1 of the Boolean algebra. This ring has the property that* a · a = a *for all* a *in* A*; rings with this property are called Boolean rings.*

Conversely, if a Boolean ring A is given, we can turn it into a Boolean algebra by defining $x \vee y := x + y + (x \cdot y)$ and $x\ \Omega\ y := x \cdot y$. Since these two constructions are inverses of each other, we can say that every Boolean ring arises from a Boolean algebra, and vice versa. Furthermore, a map $f : A \to B$ is a homomorphism of Boolean algebras if and only if it is a homomorphism of Boolean rings. The categories of Boolean rings and Boolean algebras are equivalent.

An *ideal* of the Boolean algebra A is a subset I such that for all x, y in I we have x ∨ y in I and for all ai n A we have $a\ \Omega\ x$ in I. This notion of ideal coincides with the notion of ring ideal in the Boolean ring A. An ideal I of A is called *prime* if $I \neq A$ and if $a \, \Omega \, b$ in I always implies a in I or b in I. An ideal I of A *is* called *maximal* if $I \neq A$ and if the only ideal properly containing I is A itself. These notions coincide with ring theoretic ones of prime ideal and maximal ideal in the Boolean ring A.

The dual of an *ideal* is a *filter*. A *filter* of the Boolean algebra A is a subset p such that for all x, y in p *we* have $x \, \Omega \, y$ in p and for all a in A we have $a \vee x$ in p.

Representations

It can be shown that every finite Boolean algebra is isomorphic to the Boolean algebra of all subsets of a finite set. Therefore, the number of elements of every finite Boolean algebra is a power of two.

Stone's celebrated representation theorem for Boolean algebras states that every Boolean algebra A is isomorphic to the Boolean algebra of all clopen sets in some (compact totally disconnected Hausdorff) topological space.

Axiomatics

The first axiomatisation of Boolean lattices/algebras in general was given by Alfred North Whitehead in 1898. In 1933, the American mathematician Edward Vermilye Huntington (1874–1952) set out the following elegant axiomatisation for Boolean algebra. It requires just one binary operation + and aunary functional symbol n, to be read as 'complement', which satisfy the following laws:

1. *Commutativity*: $x + y = y + x$.
2. *Associativity*: $(x + y) + z = x + (y + z)$.
3. *Huntington equation*: $n(n(x) + y) + n(n(x) + n(y)) = x$.

Herbert Robbins immediately asked: If the Huntington equation is replaced with its dual, to wit:

4. *Robbins Equation*: $n(n(x + y) + n(x + n(y))) = x$,

do (1), (2), and (4) form a basis for Boolean algebra? Calling (1), (2), and (4) a *Robbins algebra*, the question then becomes: Is every Robbins algebra a Boolean algebra? This question (which came to be known as the Robbins conjecture) remained open for decades, and became a favourite question of Alfred Tarski and his students.

In 1996, William McCune at Argonne National Laboratory, building on earlier work by Larry Wos, Steve Winker, and Bob Veroff, answered Robbins's question in the affirmative: Every Robbins algebra is a Boolean algebra. Crucial to McCune's proof was the automated reasoning program EQP he designed.

Generalisations

Removing the requirement of existence of a unit from the axioms of Boolean algebra yields "generalised Boolean algebras". Formally, a distributive lattice B *is a generalised Boolean lattice, if it has a smallest element 0 and for any elements* a *and* b *in* B *such that* a ≤ b, *there exists an element* x *such that* $a \,\Omega\, x = 0$ *and* $a \vee x = b$. *Defining* $a - b$ *as the unique* x *such that* $(a \,\Omega\, b) \vee x = a$ *and* $(a \,\Omega\, b) \,\Omega\, x = 0$, *we say that the structure* $(B, \Omega, \vee, \backslash, 0)$ *is a* generalised Boolean algebra, *while* $(B, \vee, 0)$ *is a generalised* Boolean semilattice. *Generalised Boolean lattices are exactly the ideals of Boolean lattices.*

A structure that satisfies all axioms for Boolean algebras except the two distributivity axioms is called an orthocomplemented lattice. Orthocomplemented lattices arise naturally in quantum logic as lattices of closed subspaces for separable Hilbert spaces.

Algebraic Geometry

Algebraic geometry is a branch of mathematics which combines techniques of abstract algebra, especially commutative algebra, with the language and the problems of geometry. It occupies a central place in modern mathematics and has multiple conceptual connections with such diverse fields as complex analysis, topology and number theory. Initially a study of systems of polynomial equations in several variables, the subject of algebraic geometry starts where equation solving leaves off, and it becomes even more important to understand the intrinsic properties of the totality of solutions of a system of equations, than to find some solution; this leads into some of the deepest waters in the whole of mathematics, both conceptually and in terms of technique.

The fundamental objects of study in algebraic geometry are algebraic varieties, geometric manifestations of solutions of systems of polynomial equations. Plane algebraic curves, which include lines, circles, parabolas, lemniscates, and Cassini ovals, form one of the best studied classes of algebraic varieties. A point of the plane belongs to an algebraic curve if its coordinates satisfy a given polynomial equation. Basic questions involve relative position of different curves and relations between the curves given by different equations.

Descartes's idea of coordinates is central to algebraic geometry, but it has undergone a series of remarkable transformations beginning in the early 19th century. Before then, the coordinates were assumed to be tuples of real numbers, but this changed when first complex numbers, and then elements of an arbitrary field became acceptable. Homogeneous coordinates of projective geometry offered an extension of the notion of coordinate system in a different direction, and enriched the scope of algebraic geometry. Much of the development of algebraic geometry in the 20th century occurred within an abstract algebraic framework, with increasing emphasis being placed on 'intrinsic' properties of algebraic varieties not dependent on any particular way of embedding the variety in an ambient coordinate space; this parallels developments in topology and complex geometry.

One key distinction between classical projective geometry of 19th century and modern algebraic geometry, in the form given to it by Grothendieck and Serre, is that the former is concerned with the more geometric notion of a point, while the latter emphasizes the more analytic concepts of a regular function and a regular map and extensively draws on sheaf theory. Another important difference lies in the scope of the subject. Grothendieck's idea of scheme provides the language and the tools for geometric treatment of arbitrary commutative rings and, in particular, bridges algebraic geometry with algebraic number theory. Andrew Wiles's celebrated proof of Fermat's last theorem is a vivid testament to the power of this approach. André Weil, Grothendieck, and Deligne also demonstrated that the fundamental ideas of topology of manifolds have deep analogues in algebraic geometry over finite fields.

Zeros of Simultaneous Polynomials

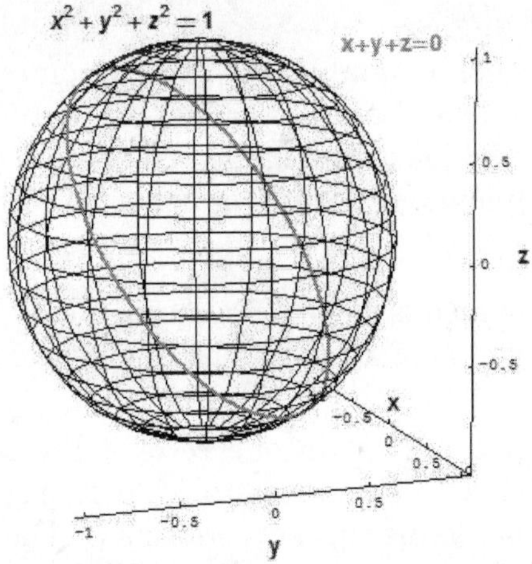

Figure 3: Sphere and slanted circle

In classical algebraic geometry, the main objects of interest are the vanishing sets of collections of polynomials, meaning the set of all points that simultaneously satisfy one or more polynomial equations. For instance, the two-dimensional sphere in three-dimensional Euclidean space R^3 could be defined as the set of all points (x,y,z) with

$$x^2 + y^2 + z^2 - 1 = 0.$$

A "slanted" circle in R^3 can be defined as the set of all points (x,y,z) which satisfy the two polynomial equations

$$x^2 + y^2 + z^2 - 1 = 0,$$

$$x + y + z = 0.$$

Affine Varieties

First we start with a field k. In classical algebraic geometry, this field was always the complex numbers C, but many of the same results are true if we assume only that k is algebraically closed. We consider the affine space of dimension n over k, denoted $A^n(k)$ (or more simply A^n, when k is clear from the context). When one fixes a coordinates system, one may identify $A^n(k)$ with k^n. The purpose of not working with k^n is to emphasize that one "forgets" the vector space structure that k^n carries.

A function $f: A^n \to A^1$ is said to be regular if it can be written as a polynomial, that is, if there is a polynomial p in $k[x_1,...,x_n]$ such that $f(t_1,...,t_n) = p(t_1,...,t_n)$ for every point $(t_1,...,t_n)$ of A^n.

Regular functions on affine n-space are thus exactly the same as polynomials over k in n variables. We will refer to the set of all regular functions on A^n as $k[A^n]$.

We say that a polynomial *vanishes* at a point if evaluating it at that point gives zero. Let S be a set of polynomials in $k[A^n]$. The *vanishing set of S* (or *vanishing locus*) is the set $V(S)$ of all points in A^n where every polynomial in S vanishes. In other words,

$$V(S) = \{(t_1,\dots,t_n) \mid \forall p \in S, p(t_1,\dots,t_n) = 0\}.$$

A subset of A^n which is $V(S)$, for some S, is called an algebraic set. The V stands for *variety* (a specific type of algebraic set to be defined below).

Given a subset U of A^n, can one recover the set of polynomials which generate it? If U is *any* subset ofA^n, define $I(U)$ to be the set of all polynomials whose vanishing set contains U. The I stands for ideal: if two polynomials f and g both vanish on U, then f+g vanishes on U, and if h is any polynomial, then hf *vanishes* on U, so $I(U)$ is always an ideal of $k[A^n]$.

Two natural questions to ask are:

- Given a subset U of A^n, when is $U = V(I(U))$?
- Given a set S of polynomials, when is $S = I(V(S))$?

The answer to the first question is provided by introducing the Zariski topology, a topology on A^n which directly reflects the algebraic structure of $k[A^n]$. Then $U = V(I(U))$ if and only if U is a Zariski-closed set. The answer to the second question is given by Hilbert's Null stellensatz. In one of its forms, it says that $I(V(S))$ is the prime radical of the ideal generated by S. In more abstract language, there is a Galois connection, giving rise to two closure operators; they can be identified, and naturally play a basic role in the theory; the example is elaborated at Galois connection.

For various reasons we may not always want to work with the entire ideal corresponding to an algebraic set U. Hilbert's basis theorem implies that ideals in $k[A^n]$ are always finitely generated.

An algebraic set is called irreducible if it cannot be written as the union of two smaller algebraic sets. An irreducible algebraic set is also called a variety. It turns out that an algebraic set is a variety if and only if the polynomials defining it generate a prime ideal of the polynomial ring.

Regular Functions

Just as continuous functions are the natural maps on topological spaces and smooth functions are the natural maps on differentiable manifolds, there is a natural class of functions on an algebraic set, called regular functions. A regular function on an algebraic set V *contained in A^n is defined to be the restriction of a regular function on A^n, in the sense we defined above.*

It may seem unnaturally restrictive to require that a regular function always extend to the ambient space, but it is very similar to the situation in a normal topological space, where the Tietze extension theorem guarantees that a continuous function on a closed subset always extends to the ambient topological space.

Just as with the regular functions on affine space, the regular functions on V *form a ring, which we denote by* k[V]*. This ring is called the coordinate ring of* V.

Since regular functions on V come from regular functions on A^n, there should be a relationship between their coordinate rings. Specifically, to get a function in k[V] *we took a function in* k[A^n]*, and we said that it was the same as another function if they gave the same* *values when evaluated on* V. *This is the same as saying that their difference is zero on V. From this we can see that* k[V] *is the quotient* $k[A^n]/I(V)$.

The Category of Affine Varieties

Using regular functions from an affine variety to A^1*, we can define regular functions from one affine variety to another. First we will define a regular function from a variety into affine space: Let* V *be a variety contained in* A^n. *Choose* m *regular functions on* V, *and call them* $f_1, ..., f_m$. *We define a regular function* f *from* V *to* A^m *by letting* $f(t_1, ..., t_n) = (f_1, ..., f_m)$. *In other words, each* f_i *determines one coordinate of the range of* f.

If V' *is a variety contained in* A^m*, we say that* f *is a regular function from* V *to* V' *if the range of* f *is contained in* V'.

This makes the collection of all affine varieties into a category, where the objects are affine varieties and the morphisms are regular maps. The following theorem characterises the category of affine varieties:

The category of affine varieties is the opposite category to the category of finitely generated integral k-algebras and their homomorphisms.

Projective Space

Consider the variety $V(y - x^2)$. If we draw it, we get a parabola. As x increases, the slope of the line from the origin to the point (x, x^2) becomes larger and larger. As x decreases, the slope of the same line becomes smaller and smaller.

Compare this to the variety $V(y - x^3)$. This is a cubic equation. As x increases, the slope of the line from the origin to the point (x, x^3) becomes larger and larger just as before. But unlike before, as x decreases, the slope of the same line again becomes larger and larger. So the behaviour "at infinity" of $V(y - x^3)$ is different from the behaviour "at infinity" of $V(y - x^2)$. It is, however, difficult to make the concept of "at infinity" meaningful, if we restrict to working in affine space.

The remedy to this is to work in projective space. Projective space has properties analogous to those of a compact Hausdorff space. Among other things, it lets us make precise the notion of "at infinity" by including extra points. The behaviour of a variety at those extra points then gives us more information about it. As it turns out, $V(y - x^3)$ has a singularity at one of those extra points, but $V(y - x^2)$ is smooth.

While projective geometry was originally established on a synthetic foundation, the use of homogeneous coordinates allowed the introduction of algebraic techniques. Furthermore, the introduction of projective techniques made many theorems in algebraic geometry simpler and sharper: For example, Bézout's theorem on the number of intersection points between two varieties can be stated in its sharpest form only in projective space. For this reason, projective space plays a fundamental role in algebraic geometry.

The Modern Viewpoint

The modern approaches to algebraic geometry redefine and effectively extend the range of basic objects in various levels of generality to schemes, formal schemes, ind-schemes, algebraic spaces, algebraic stacks and so on. The need for this arises already from the useful ideas within theory of varieties, e.g. the formal functions of Zariski can be accommodated by introducing nilpotent elements in structure rings; considering spaces of loops and arcs, constructing quotients by group actions and developing formal grounds for natural intersection theory and deformation theory lead to some of the further extensions.

Most remarkably, in late 1950s, algebraic varieties were subsumed into Alexander Grothendieck's concept of a scheme. Their local objects are affine schemes or prime spectra which are locally ringed spaces which form a category which is antiequivalent to the category of commutative unital rings, extending the duality between the category of affine algebraic varieties over a field k, and the category of finitely generated reduced k-algebras.

The gluing is along Zariski topology; one can glue within the category of locally ringed spaces, but also, using the Yoneda embedding, within the more abstract category of presheaves of sets over the category of affine schemes. The Zariski topology in the set theoretic sense is then replaced by a Zariski topology in the sense of Grothendieck topology. Grothendieck introduced Grothendieck topologies having in mind more exotic but geometrically finer and more sensitive examples than the crude Zariski topology, namely the étale topology, and the two flat Grothendieck topologies: ffpf and fpqc; nowadays some other examples became prominent including Nisnevich topology. Sheaves can be furthermore generalised to stacks in the sense of Grothendieck, usually with some additional representability conditions leading to Artin stacks and, even finer, Deligne-Mumford stacks, both often called algebraic stacks.

Sometimes other algebraic sites replace the category of affine schemes. For example, Nikolai Durovhas introduced commutative algebraic monads as a generalisation of local objects in a generalised algebraic geometry. Versions of a tropical geometry, of an absolute geometry over a field of one element and an algebraic analogue of Arakelov's geometry were realised in this setup.

Another formal generalisation is possible to Universal algebraic geometry in which every variety of algebra has its own algebraic geometry. The term variety of algebra should not be confused with algebraic variety.

The language of schemes, stacks and generalisations has proved to be a valuable way of dealing with geometric concepts and became cornerstones of modern algebraic geometry.

Algebraic stacks can be further generalised and for many practical questions like deformation theory and intersection theory, this is often the most natural approach. One can extend the Grothendieck site of affine schemes to a higher categorical site of derived affine schemes, by replacing the commutative rings with an infinity category of differential graded commutative algebras, or of simplicial commutative rings or a similar category with an appropriate variant of a Grothendieck topology. One can also replace presheaves of sets by presheaves of simplicial sets (or of infinity groupoids).

Then, in presence of an appropriate homotopic machinery one can develop a notion of derived stack as such a presheaf on the infinity category of derived affine schemes, which is satifsying certain infinite categorical version of a sheaf axiom (and to be algebraic, inductively a sequence of representability conditions). Quillen model categories, Segal categories and quasicategories are some of the most often used tools to formalise this yielding the derived algebraic geometry, introduced by the school of Carlos Simpson, including Andre Hirschowitz, Bertrand Toën, Gabrielle Vezzosi, Michel Vaquié and others; and developed further by Jacob Lurie, Bertrand Toën, and Gabrielle Vezzosi. Another (noncommutative) version of derived algebraic geometry, using A-infinity categories has been developed from early 1990-s by Maxim Kontsevich and followers.

History

Prehistory: Before the 19th Century: Some of the roots of algebraic geometry date back to the work of the Hellenistic Greeks from the 5th century BC. The Delian problem, for instance, was to construct a length x so that the cube of side x *contained* the same volume as the rectangular box a^2b for given sides a and b. Menechmus (circa 350 BC) considered the problem geometrically by intersecting the pair of plane conics $ay = x^2$ and $xy = ab$.

The later work, in the 3rd century BC, of Archimedes and Apollonius studied more systematically problems on conic sections, and also involved the use of coordinates. The Arab mathematicians were able to solve by purely algebraic means certain cubic equations, and then to interpret the results geometrically. This was done, for instance, by Ibn al-Haytham in the 10th century AD. Subsequently, Persian mathematician Omar Khayyám (born 1048 A.D.) discovered the general method of solving cubic equations by intersecting a parabola with a circle. Each of these early developments in algebraic geometry dealt with questions of finding and describing the intersections of algebraic curves.

Such techniques of applying geometrical constructions to algebraic problems were also adopted by a number of Renaissance mathematicians such as Gerolamo Cardano and Niccolò Fontana "Tartaglia" on their studies of the cubic equation. The geometrical approach to construction problems, rather than the algebraic one, was favoured by most 16th and 17th century mathematicians, notably Blaise Pascal who argued against the use of algebraic and analytical methods in geometry. The French mathematicians Franciscus

Vieta and later René Descartes and Pierre de Fermat revolutionised the conventional way of thinking about construction problems through the introduction of coordinate geometry. They were interested primarily in the properties of *algebraic curves*, such as those defined by Diophantine equations (in the case of Fermat), and the algebraic reformulation of the classical Greek works on conics and cubics (in the case of Descartes).

During the same period, Blaise Pascal and Gérard Desargues approached geometry from a different perspective, developing the synthetic notions of projective geometry. Pascal and Desargues also studied curves, but from the purely geometrical point of view: the analog of the Greek *ruler and compass construction*. Ultimately, the analytic geometry of Descartes and Fermat won out, for it supplied the 18th century mathematicians with concrete quantitative tools needed to study physical problems using the new calculus of Newton and Leibniz. However, by the end of the 18th century, most of the algebraic character of coordinate geometry was subsumed by the *calculus of Infinitesimals* of Lagrange and Euler.

Nineteenth and Early 20th Century

It took the simultaneous 19th century developments of non-Euclidean geometry and Abelian integralsin order to bring the old algebraic ideas back into the geometrical fold. The first of these new developments was seized up by Edmond Laguerre and Arthur Cayley, who attempted to ascertain the generalised metric properties of projective space. Cayley introduced the idea of *homogeneous polynomial forms*, and more specifically quadratic forms, on projective space. Subsequently, Felix Klein studied projective geometry (along with other sorts of geometry) from the viewpoint that the geometry on a space is encoded in a certain class of transformations on the space.

By the end of the 19th century, projective geometres were studying more general kinds of transformations on figures in projective space. Rather than the projective linear transformations which were normally regarded as giving the fundamental Kleinian geometry on projective space, they concerned themselves also with the higher degree birational transformations. This weaker notion of congruence would later lead members of the 20th century Italian school of algebraic geometry to classify algebraic surfaces up to birational isomorphism.

The second early 19th century development, that of Abelian integrals, would lead Bernhard Riemann to the development of Riemann surfaces.

Twentieth Century

B. L. van der Waerden, Oscar Zariski, André Weil and others attempted to develop a rigorous foundation for algebraic geometry based on contemporary commutative algebra, including valuation theory and the theory of ideals.

In the 1950s and 1960s Jean-Pierre Serre and Alexander Grothendieck recast the foundations making use of sheaf theory. Later, from about 1960, and largely spearheaded by Grothendieck, the idea of schemes was worked out, in conjunction with a very refined apparatus of homological techniques. After a decade of rapid development the field stabilised in the 1970s, and new applications were made, both to number theory and to more classical geometric questions on algebraic varieties, singularities and moduli.

An important class of varieties, not easily understood directly from their defining equations, are the abelian varieties, which are the projective varieties whose points form an abelian group. The prototypical examples are the elliptic curves, which have a rich theory. They were instrumental in the proof of Fermat's last theorem and are also used in elliptic curve cryptography.

While much of algebraic geometry is concerned with abstract and general statements about varieties, methods for effective computation with concretely-given polynomials have also been developed. The most important is the technique of Grφbner bases which is employed in all computer algebra systems. Based on these methods, several solvers may compute all the solutions of a system of polynomial equations whose associated variety has dimension zero and thus consists in a finite number of points.

Applications

Algebraic geometry now finds application in statistics, control theory, robotics, error-correcting codes, phylogenetics and geometric modelling. There are also connections to string theory, game theory, graph matchings, solitons and integer programming. Google Scholar lists hundreds of more studies on algebraic geometry in biology, chemistry, economics, physics and of course other areas of mathematics.

Algebraic Combinatorics

Algebraic combinatorics is an area of mathematics that employs methods of abstract algebra, notably group theory and representation theory, in various combinatorial contexts and, conversely, applies combinatorial techniques to problems in algebra.

Algebraic Combinatorics is one of the youngest combinatorial disciplines. Thus, in the preface to his Combinatorial *Theory*, published in 1979, Martin Aigner wrote about "growing consensus that combinatorics should be divided into three parts" (*Enumeration, Order theory, Configurations*), without even mentioning algebraic combinatorics by name. The book *Algebraic Combinatorics* by Bannai and Ito was published in 1983.

Through the early or mid 1990s, typical combinatorial objects of interest in algebraic combinatorics either admitted a lot of symmetries (association schemes, strongly regular graphs, posets with agroup action) or possessed a rich algebraic structure, frequently of representation theoretic origin (symmetric functions, Young tableaux). This period is reflected in the area 05E, *Algebraic combinatorics*, of the AMS Mathematics Subject Classification, introduced in 1991.

However, within the last decade or so, algebraic combinatorics came to be seen more expansively as an area of mathematics where the interaction of combinatorial and algebraic methods is particularly strong and significant. Thus the combinatorial topics may be enumerative in nature or involve matroids, polytopes, partially ordered sets, or finite geometries.

On the algebraic side, besides group and representation theory, lattice theory and commutative algebra are common. One of the fastest developing subfields within algebraic combinatorics is combinatorial commutative algebra. Journal of Algebraic Combinatorics, published by Springer-Verlag, is an international journal intended as a forum for papers in the field.

Chapter 4

Algebraic Theories

Algebraic Number Theory

Algebraic number theory is a major branch of number theory which studies algebraic structures related to algebraic integers. This is generally accomplished by considering a ring of algebraic integers O in an algebraic number field $K/\mathbb{Q}$, and studying their algebraic properties such as factorisation, the behaviour of ideals, and field extensions. In this setting, the familiar features of the integers—such as unique factorisation—need not hold. The virtue of the primary machinery employed—Galois theory, group cohomology, group representations, and *L*-functions—is that it allows one to deal with new phenomena and yet partially recover the behaviour of the usual integers.

Basic Notions

Unique Factorisation and the Ideal Class Group: One of the first properties of Z that can fail in the ring of integers *O* of an algebraic number field *K* is that of the unique factorisation of integers into prime numbers. The prime numbers in Z are generalised to irreducible elements in *O*, and though the unique factorisation of elements of *O* into irreducible elements may hold in some cases (such as for the Gaussian integers Z[i]), it may also fail, as in the case of Z[Φ-5] where

$$6 = 2 \cdot 3 = (1+\sqrt{-5}) \cdot (1-\sqrt{-5}).$$

The ideal class group of *O* is a measure of how much unique factorisation of elements fails; in particular, the ideal class group is trivial if, and only if, *O* is a unique factorisation domain.

Factoring Prime Ideals in Extensions

Unique factorisation can be partially recovered for *O* in that it has the property of unique factorisation of ideals into prime ideals (i.e. it is a Dedekind domain). This makes the study of the prime ideals in O *particularly* important. This is another area where things change from Z to O: the prime numbers, which generate prime ideals of Z (in fact, every single prime ideal of Z is of the form (p):=pZ for some prime number p,) may no longer generate prime ideals in O. For example, in the ring of Gaussian integers, the ideal 2Z[i] is no longer a prime ideal; in fact

$$2Z[i] = ((1+i)Z[i])^2.$$

On the other hand, the ideal 3Z[i] is a prime ideal. The complete answer for the Gaussian integers is obtained by using a theorem of Fermat's, with the result being that for an odd prime number p

$$pZ[i] \text{ is a prime ideal if } p \equiv 3 (\text{mod} 4)$$

$$pZ[i] \text{ is not a prime ideal if } p \equiv 1 (\text{mod} 4).$$

Generalising this simple result to more general rings of integers is a basic problem in algebraic number theory. Class field theory accomplishes this goal when K is an abelian extension of Q (i.e. a Galois extension with abelian Galois group).

Primes and Places

An important generalisation of the notion of prime ideal in O is obtained by passing from the so-called ideal-*theoretic* approach to the so-called *valuation-theoretic* approach. The relation between the two approaches arises as follows. In addition to the usual absolute value function $|\cdot| : Q \to R$, there are absolute value functions $|\cdot|_p : Q \to R$ defined for each prime number p in Z, called p-adic absolute values. Ostrowski's theorem states that these are all possible absolute value functions on Q (up to equivalence). This suggests that the usual absolute value could be considered as another prime. More generally, a prime of an algebraic number field K (also called a place) is an equivalence class of absolute values on K. The primes in K are of two sorts: $\mathfrak{p}$ -adic absolute values like $|\cdot|_{\mathfrak{p}}$, one for each prime ideal $\mathfrak{p}$ of O, and absolute values like $|\cdot|$ obtained by considering K as a subset of the complex numbers in various possible ways and using the absolute value $|\cdot| : C \to R$. A prime of the first kind is called a finite prime (or finite place) and one of the second kind is called an infinite prime (or infinite place). Thus, the set of primes of Q is generally denoted { 2, 3, 5, 7, ..., ∞ }, and the usual absolute value on Q is often denoted $|\cdot|_\infty$ in this context.

The set of infinite primes of K can be described explicitly in terms of the embeddings $K \to C$ (i.e. the non-zero ring homomorphisms from K to C). Specifically, the set of embeddings can be split up into two disjoint subsets, those whose image is contained in R, and the rest. To each embedding $\sigma : K \rightarrow$R, there corresponds a unique prime of K coming from the absolute value obtained by composing σ with the usual absolute value on R; a prime arising in this fashion is called a real prime (or real place).

To an embedding $\tau : K \rightarrow$ C whose image is *not* contained in R, one can construct a distinct embedding τ, called the *conjugate embedding*, by composing τ with the complex conjugation map C$\rightarrow$ C. Given such a pair of embeddings τ and τ, there corresponds a unique prime of K again obtained by composing τ with the usual absolute value (composing τ instead gives the same absolute value function since $|z| = |z|$ for any complex number z, where z denotes the complex conjugate of z). Such a prime is called a complex prime (or complex place). The description of the set of infinite primes is then as follows: each infinite prime corresponds either to a unique embedding $\sigma : K \rightarrow$ R, or a pair of conjugate embeddings $\tau, \tau : K \rightarrow$ C. The number of real (respectively, complex) primes is often denoted r_1 (respectively, r_2). Then, the total number of embeddings $K \rightarrow$ C is r_1+2r_2 (which, in fact, equals the degree of the extension K/Q).

Units

The fundamental theorem of arithmetic describes the multiplicative structure of Z. It states that every non-zero integer can be written (essentially) uniquely as a product of prime powers and ±1. The unique factorisation of ideals in the ring O recovers part of this description, but fails to address the factor ±1.

The integers 1 and -1 are the invertible elements (i.e. units) of Z. More generally, the invertible elements in O form a group under multiplication called the unit group of O, denoted $O^{\times}$. This group can be much larger than the cyclic group of order 2 formed by the units of Z. Dirichlet's unit theorem describes the abstract structure of the unit group as an abelian group. A more precise statement giving the structure of $O^{\times}$ Δ_Z Q as a Galois module for the Galois group of K/Q is also possible. The size of the unit group, and its lattice structure give important numerical information about O, as can be seen in the class number formula.

Local Fields

Completing a number field K at a place w gives a complete field. If the valuation is archimedean, one gets R or C, if it is non-archimedean and lies over a prime p of the rationals, one gets a finite extension Kw / Q_p: a complete, discrete valued field with finite residue field. This process simplifies the arithmetic of the field and allows the local study of problems. For example the Kronecker–Weber theorem can be deduced easily from the analogous local statement.

The philosophy behind the study of local fields is largely motivated by geometric methods. In algebraic geometry, it is common to study varieties locally at a point by localising to a maximal ideal. Global information can then be recovered by gluing together local data. This spirit is adopted in algebraic number theory. Given a prime in the ring of algebraic integers in a number field, it is desirable to study the field locally at that prime. Therefore one localises the ring of algebraic integers to that prime and then completes the fraction field much in the spirit of geometry.

Major Results

Finiteness of the Class Group: One of the classical results in algebraic number theory is that the ideal class group of an algebraic number field K is finite. The order of the class group is called the class number, and is often denoted by the letter h.

Dirichlet's Unit Theorem

Dirichlet's unit theorem provides a description of the structure of the multiplicative group of units $O^\times$ of the ring of integers O. Specifically, it states that $O^\times$ is isomorphic to $G \times Z^r$, where G is the finite cyclic group consisting of all the roots of unity in O, and $r=r_1+r_2-1$ (where r_1 (respectively, r_2) denotes the number of real embeddings (respectively, pairs of conjugate non-real embeddings) of K). In other words, $O^\times$ is a finitely generated abelian group of rank $r_1 + r_2 - 1$ whose torsion consists of the roots of unity in O.

Fundamental Theorem of Arithmetic

In number theory, the fundamental theorem of arithmetic (or the unique-prime-factorisation theorem) states that any integer greater than 1 can be written as a *unique* product (up to ordering of the factors) of prime numbers. For example,

$$6936 = 2^3 \times 3^1 \times 17^2,$$

$$1200 = 2^4 \times 3^1 \times 5^2,$$

are two numbers satisfying the hypothesis of the theorem that can be written as the product of prime numbers.

Proof of existence of a prime factorisation is straightforward; proof of *uniqueness* is more challenging. Some proofs use the fact that if a prime number p divides the product of two natural numbers a and b, then p divides either a or b, a statement known as Euclid's lemma. Since multiplication on the integers is both commutative and associative, it does not matter in what way we write a number greater than 1 as the product of primes; it is common to write the (prime) factors in the order of smallest to largest.

Some natural extensions of the hypothesis of this theorem allow *any non-zero integer* to be expressed as the product of "prime numbers" and "invertibles". For example, 1 and "1 are allowed to be factors of such representations (although they are not considered to be prime). In this way, one can extend the Fundamental Theorem of Arithmetic to any Euclidean domain or principal ideal domain bearing in mind certain alterations to the hypothesis of the theorem. A ring in which the Fundamental Theorem of Arithmetic holds is called a unique factorisation domain.

Many authors assume 1 to be a natural number that has no prime factorisation. Thus Theorem 1 of Hardy & Wright (1979) takes the form, "Every positive integer, except 1, is a product of one or more primes," and Theorem 2 (their "Fundamental") asserts uniqueness. By convention, the number 1 is not itself prime, but since it is the product of no numbers, it is often convenient to include it in the theorem by the empty product rule.

Applications

The fundamental theorem of arithmetic establishes the importance of prime numbers. Prime numbers are the basic building blocks of any positive integer, in the sense that each positive integer can be constructed from the product of primes with one unique construction. Finding the prime factorisation of an integer allows derivation of all its divisors, both prime and non-prime. This property of uniqueness is powerful, as the factorisation can be thought of as a key for a number. This is why it can be used to build powerful encryption techniques such as RSA.

For example, the above factorisation of 6936 shows that any positive divisor of 6936 must have the form $2^a \times 3^b \times 17^c$*, where* a *takes one of the 4 values in {0, 1, 2, 3}, where* b *takes one of the 2 values in {0, 1}, and where* c *takes one of the 3 values in {0, 1, 2}. Multiplying the numbers of independent options together produces a total of* $4 \times 2 \times 3 = 24$ *positive divisors.*

Once the prime factorisations of two numbers are known, their greatest common divisor and least common multiple can be found quickly. For instance, from the above it is shown that the greatest common divisor of 6936 and 1200 is $2^3 \times 3 = 24$. However, if the prime factorisations are not known, the use of the Euclidean algorithm generally requires much less calculation than factoring the two numbers. The fundamental theorem ensures that additive and multiplicative arithmetic functions are completely determined by their values on the powers of prime numbers.

It may be important to note that Egyptians like Ahmes used earlier practical aspects of the factoring, such as the least common multiple, allowing a long tradition to develop, as formalised by Euclid, and rigorously proven by Carl Friedrich Gauss. Although at first sight the theorem seems 'obvious', it does not hold in more general number systems, including many rings of algebraic integers. This was first pointed out by Ernst Kummer in 1843, in his work on Fermat's Last Theorem. The recognition of this failure is one of the earliest developments in algebraic number theory.

Euclid's Proof

The proof consists of two steps. In the first step every number is shown to be a product of zero or more primes. In the second step, the proof shows that any two representations may be unified into a single representation.

Non-prime Composite Numbers

Proof by contradiction via minimal counter example: Suppose there were positive integers which cannot be written as a product of primes. Then by the well-ordering principle there must be a smallest such number, n. Because of the empty-product convention above, n cannot be 1. Also, n cannot be a prime number, since any prime number is a product of a single prime: itself. So n must be a composite number. Thus

$$n = a \cdot b,$$

where both a and b are positive integers smaller than n. Since n is the smallest number which cannot be written as a product of primes, both a and b *can* be written as products of primes. But then

$$n = a \cdot b,$$

can be written as a product of primes as well, simply by combining the factorisations of a and b; but this contradicts our supposition. Hence, all positive integers can be written as a product of primes.

Proof of Uniquencss

The key step in proving uniqueness is Euclid's proposition 30 of book 7 (known as Euclid's lemma), which states that, for any prime number p and any natural numbers a, b: if p divides ab then p dividesa or p divides b.

This may be proved as follows:

- Suppose that a prime p divides ab (where a, b are natural numbers) but does not divide a. We must prove that p divides b.
- Since p does not divide a, and p is prime, the greatest common divisor of p and a is 1.
- By Bézout's identity, it follows that for some integers x, y (possibly negative), $\mathrm{px + ay = 1}$.
- Multiplying both sides by b, $\mathrm{pxb + aby = b}$.
- Since p divides ab, p divides (ab)y
- Since p divides both summands on the left, p divides b.

A proof of the uniqueness of the prime factorisation of a given integer proceeds as follows. Let s be the smallest natural number that can be written as (at least) two different products of prime numbers. Denote these two factorisations of s as $p_1 \cdots p_m$ and $q_1 \cdots q_n$, such that $s = p_1 p_2 \cdots p_m = q_1 q_2 \cdots q_n$. Both q1 and $q_2 \cdots q_n$ must have unique prime factorisations (since both are smaller than s). By Euclid's proposition either p_1 divides q_1, or p_1 divides $q_2 \cdots q_n$. Therefore $p_1 = q_j$ (for some j). But by removing p_1 and q_j from the initial equivalence we have a smaller integer factorisable in two ways, contradicting our initial assumption. Therefore there can be no such s, and all natural numbers have a unique prime factorisation.

Alternate Proof of Uniqueness

Assume that s is the least integer that can be written as (at least) two different products of prime numbers. Denote these two factorisations of s as $p_1 \cdots p_m$ and $q_1 \cdots q_n$, such that $s = p_1 p_2 \cdots p_m = q_1 q_2 \cdots q_n$.

No p_i (with $1 \le i \le m$) can be equal to any q_j (with $1 \le j \le n$), as there would otherwise be a smaller integer factorisable in two ways (by removing prime factors common in both products), violating the above assumption. Now it can be assumed without loss of generality that p_1 is a prime factor smaller than any q_j (with $1 \ne j \ne n$). Note that q_1 divides s / p_1. Since s / p_1 is smaller than s, it has an unique factorisation $\mathrm{p_2 p_3 \cdots p_m}$. This is impossible.

Canonical Representation of a Positive Integer

Since the primes into which an integer can be factored need not be distinct, it follows from the Fundamental Theorem of Arithmetic that each integer $n \leq 2$ can be represented as a product of prime powers

$$n = p_1^{\alpha_1} p_2^{\alpha_2} \cdots p_k^{\alpha_k}$$

that is,

$$n = \prod_{i=1}^{k} p_i^{\alpha_i}$$

where α_i are positive integers and p_i are primes satisfying

$$p_1 < p_2 < \cdots < p_k .$$

This representation is called the canonical representation of *n*, or the standard form of *n*. Thus, $2^2 \cdot 3$, $2^4 \cdot 3^4$, and $2^2 \cdot 5 \cdot 11$ are the canonical representations of 12, 1296, and 220 in that order. Each positive integer has exactly one such representation.

If, in a given problem, only one number is represented in this way, we usually require α_i to be positive for each *i*, as above. However, for notational convenience when two or more numbers are involved, we sometimes allow some of the exponents to be zero.

If $\alpha_1 = 0$, for example, the prime p_1 simply does not occur in the canonical representation of *n*. This device makes it possible to write the canonical representation of any two positive integers so that they *appear* to involve the same prime factors even though they may, in fact, fail to have any nontrivial common factor. For example, we could write $12 = 2^2 \cdot 3 \cdot 5^0$ and $20 = 2^2 \cdot 3^0 \cdot 5$. And one could even write $1 = 2^0 \cdot 3^0 \cdot 5^0$.

Generalisations

The fundamental theorem of arithmetic generalises to various contexts; for example in the context ofring theory, where the field of algebraic number theory develops. A ring is said to be a unique factorisation domain if the Fundamental theorem of arithmetic (for non-zero elements) holds there. For example, any Euclidean domain or principal ideal domain is necessarily a unique factorisation domain. Specifically, a field is trivially a unique factorisation domain.

Fundamental Theorem of Algebra

The fundamental theorem of algebra states that every non-constant single-variable polynomial with complex coefficients has at least one complex root. Equivalently, the field of complex numbers is algebraically closed. Sometimes, this theorem is stated as: every non-zero single-variable polynomial with complex coefficients has exactly as many complex roots as its degree, if each root is counted up to its multiplicity. Although this at first appears to be a stronger statement, it is a direct consequence of the other form of the theorem, through the use of successive polynomial division by linear factors.

In spite of its name, there is no purely algebraic proof of the theorem, since any proof must use the completeness of the reals (or some other equivalent formulation of completeness), which is not an algebraic concept. Additionally, it is not fundamental for modern algebra; its name was given at a time when the study of algebra was mainly concerned with the solutions of polynomial equations with real or complex coefficients.

History

Peter Rothe (Petrus Roth), in his book Arithmetica Philosophica (published in 1608), wrote that a polynomial equation of degree n (with real coefficients) may have n solutions. Albert Girard, in his book L'invention nouvelle en l'Algèbre (published in 1629), asserted that a polynomial equation of degree n has n solutions, but he did not state that they had to be real numbers. Furthermore, he added that his assertion holds "unless the equation is incomplete", by which he meant that no coefficient is equal to 0. However, when he explains in detail what he means, it is clear that he actually believes that his assertion is always true; for instance, he shows that the equation $x^4 = 4x - 3$, although incomplete, has four solutions (counting multiplicities): 1 (twice), -1 + iΦ2, and -1 - iΦ2.

As will be mentioned again below, it follows from the fundamental theorem of algebra that every non-constant polynomial with real coefficients can be written as a product of polynomials with real coefficients whose degree is either 1 or 2. However, in 1702 Leibniz said that no polynomial of the type $x^4 + a^4$ *(with* a *real and distinct from 0) can be written in such a way. Later, Nikolaus Bernoulli made the same assertion concerning the polynomial* $x^4 - 4x^3 + 2x^2 + 4x + 4$*, but he got a letter from Euler in 1742 in which he was told that his polynomial happened to be equal to*

$$(x^2 - (2+\alpha)x + 1 + \sqrt{7} + \alpha)(x^2 - (2-\alpha)x + 1 + \sqrt{7} - \alpha),$$

where α is the square root of 4 + 2Φ7. Also, Euler mentioned that

$$x^4 + a^4 = (x^2 + a\sqrt{2}\cdot x + a^2)(x^2 - a\sqrt{2}\cdot x + a^2).$$

A first attempt at proving the theorem was made by d'Alembert in 1746, but his proof was incomplete. Among other problems, it assumed implicitly a theorem (now known as Puiseux's theorem) which would not be proved until more than a century later, and furthermore the proof assumed the fundamental theorem of algebra. Other attempts were made by Euler (1749), de Foncenex (1759), Lagrange (1772), and Laplace (1795). These last four attempts assumed implicitly Girard's assertion; to be more precise, the existence of solutions was assumed and all that remained to be proved was that their form was $a + bi$ for some real numbers a and b. In modern terms, Euler, de Foncenex, Lagrange, and Laplace were assuming the existence of a splitting field of the polynomial $p(z)$.

At the end of the 18th century, two new proofs were published which did not assume the existence of roots. One of them, due to James Wood and mainly algebraic, was published in 1798 and it was totally ignored. Wood's proof had an algebraic gap. The other one was published by Gauss in 1799 and it was mainly geometric, but it had a topological gap, filled by Alexander Ostrowski in 1920, as discussed in Smale 1981 (Smale writes, "...I wish to point out what an immense gap Gauss' proof contained.

It is a subtle point even today that a real algebraic plane curve cannot enter a disk without leaving. In fact even though Gauss redid this proof 50 years later, the gap remained. It was not until 1920 that Gauss' proof was completed. In the reference Gauss, A. Ostrowski has a paper which does this and gives an excellent discussion of the problem as well..."). A rigorous proof was published by Argand in 1806; it was here that, for the first time, the fundamental theorem of algebra was stated for polynomials with complex coefficients, rather than just real coefficients. Gauss produced two other proofs in 1816 and another version of his original proof in 1849.

The first textbook containing a proof of the theorem was Cauchy's *Cours d'analyse de l'École Royale Polytechnique* (1821). It contained Argand's proof, although Argand is not credited for it.

None of the proofs mentioned so far is constructive. It was Weierstrass who raised for the first time, in the middle of the 19th century, the problem of finding a constructive proof of the fundamental theorem of algebra. He presented his solution, that amounts in modern terms to a combination of the Durand–Kerner method with the homotopy continuation principle, in 1891. Another proof of this kind was obtained by Hellmuth Kneser in 1940 and simplified by his son Martin Kneser in 1981.

Without using countable choice, it is not possible to constructively prove the fundamental theorem of algebra for complex numbers based on the Dedekind real numbers (which are not constructively equivalent to the Cauchy real numbers without countable choice). However, Fred Richman proved a reformulated version of the theorem that does work.

Proofs

All proofs below involve some analysis, at the very least the concept of continuity of real or complex functions. Some also use differentiable or even analytic functions. This fact has led some to remark that the Fundamental Theorem of Algebra is neither fundamental, nor a theorem of algebra.

Some proofs of the theorem only prove that any non-constant polynomial with real coefficients has some complex root. This is enough to establish the theorem in the general case because, given a non-constant polynomial p(z) *with complex coefficients, the polynomial*

$$q(z) = p(z)\overline{p(\overline{z})}$$

has only real coefficients and, if z is a zero of $q(z)$, then either z or its conjugate is a root of $p(z)$.

A large number of non-algebraic proofs of the theorem use the fact (sometimes called "growth lemma") that an n-th degree polynomial function $p(z)$ whose dominant coefficient is 1 behaves like z^n when $|z|$ is large enough. A more precise statement is: there is some positive real number R such that:

$$\tfrac{1}{2}|z^n| < |p(z)| < \tfrac{3}{2}|z^n|$$

when $|z| > R$.

Complex-analytic Proofs

Find a closed disk D of radius r centred at the origin such that $|p(z)| > |p(0)|$ whenever $|z| \geq r$. The minimum of $|p(z)|$ on D, which must exist since D is compact, is therefore achieved at some point z_0 in the interior of D, but not at any point of its boundary. The minimum modulus principle implies then that $p(z_0) = 0$. In other words, z_0 is a zero of $p(z)$.

A variation of this proof that does not require the use of the minimum modulus principle (most proofs of which in turn require the use of Cauchy's integral theorem or some of its consequences) is based on the observation that for the special case of a polynomial function, the minimum modulus principle can be proved directly using elementary arguments. More precisely, if we assume by contradiction that $a := p(z_0) \neq 0$, then, expanding $p(z)$ in powers of $z - z_0$ we can write

$$p(z) = a + c_k(z - z_0)^k + c_{k+1}(z - z_0)^{k+1} + \ldots + c_n(z - z_0)^n.$$

Here, the c_j's are simply the coefficients of the polynomial $z \to p(z + z_0)$, and we let k be the index of the first coefficient following the constant term that is non-zero. But now z *sufficiently* close to z_0 this has behaviour asymptotically similar to the simpler polynomial $q(z) = a + c_k(z - z_0)^k$, in the sense that (as is easy to check) the function $\left|\frac{p(z) - q(z)}{(z - z_0)^{k+1}}\right|$ is bounded by some positive constant M in some neighbourhood of z_0. Therefore if we defines $\theta_0 = (\arg(a) + \pi - \arg(c_k)) / k$ and let $z = z_0 + re^{i\theta_0}$, then for any sufficiently small positive number r (so that the bound M mentioned above holds),

$$|p(z)| \quad < |q(z)| + r^{k+1}\left|\frac{p(z) - q(z)}{r^{k+1}}\right|$$

$$[.2em] \quad \leq \left|a + (-1)c_k r^k e^{i(\arg(a) - \arg(c_k))}\right| + Mr^{k+1}$$

$$[.5em] \quad = |a| - |c_k| r^k + Mr^{k+1}.$$

When r is sufficiently close to 0 this upper bound for $|p(z)|$ is strictly smaller than $|a|$, in contradiction to the definition of z_0. (Geometrically, we have found an explicit direction θ_0 such that if one approaches z_0 from that direction one can obtain values $p(z)$ smaller in absolute value than $|p(z_0)|$.)

Another analytic proof can be obtained along this line of thought observing that, since $|p(z)| > |p(0)|$ outside D, the minimum of $|p(z)|$ on the whole complex plane is achieved at z_0. If $|p(z_0)| > 0$, then $1/p$is a bounded holomorphic function in the entire complex plane since, for each complex number z, $|1/p(z)| \leq |1/p(z_0)|$. Applying Liouville's theorem, which states that a bounded entire function must be constant, this would imply that $1/p$ is constant and therefore that p is constant. This gives a contradiction, and hence $p(z_0) = 0$.

Yet another analytic proof uses the argument principle. Let R be a positive real number large enough so that every root of $p(z)$ has absolute value smaller than R; such a number must exist because every non-constant polynomial function of degree n has at most n zeros. For each $r > R$, consider the number

$$\frac{1}{2\pi i}\int_{c(r)} \frac{p'(z)}{p(z)} dz,$$

where $c(r)$ is the circle centred at 0 with radius r oriented counter clockwise; then the argument principle says that this number is the number N of zeros of $p(z)$ in the open ball centred at 0 with radius r, which, since $r > R$, is the total number of zeros of $p(z)$. On the other hand, the integral of n/z along $c(r)$ divided by $2\varphi i$ is equal to n. But the difference between the two numbers is

$$\frac{1}{2\pi i}\int_{c(r)}\left(\frac{p'(z)}{p(z)}-\frac{n}{z}\right)dz=\frac{1}{2\pi i}\int_{c(r)}\frac{zp'(z)-np(z)}{zp(z)}dz.$$

The numerator of the rational expression being integrated has degree at most n - 1 and the degree of the denominator is $n + 1$. Therefore, the number above tends to 0 as r tends to $+\infty$. But the number is also equal to N - n and so $N = n$.

Still another complex-analytic proof can be given by combining linear algebra with the Cauchy theorem. To establish that every complex polynomial of degree $n > 0$ has a zero, it suffices to show that every complex square matrix of size $n > 0$ has a (complex) eigen value. The proof of the latter statement is by contradiction.

Let A be a complex square matrix of size $n > 0$ and let I_n be the unit matrix of the same size. Assume A has no eigenvalues. Consider the resolvent function

$$R(z)=(zI_n-A)^{-1},$$

which is a meromorphic function on the complex plane with values in the vector space of matrices. The eigenvalues of A are precisely the poles of $R(z)$. Since, by assumption, A has no eigenvalues, the function $R(z)$ is an entire function and Cauchy's theorem implies that

$$\int_{c(r)}R(z)dz=0.$$

On the other hand, $R(z)$ expanded as a geometric series gives:

$$R(z)=z^{-1}(I_n-z^{-1}A)^{-1}=z^{-1}\sum_{k=0}^{\infty}\frac{1}{z^k}A^k.$$

This formula is valid outside the closed disc of radius $||A||$ (the operator norm of A). Let $r > ||A||$. Then

$$\int_{c(r)}R(z)dz=\sum_{k=0}^{\infty}\int_{c(r)}\frac{dz}{z^{k+1}}A^k=2\pi iI_n$$

(in which only the summand $k = 0$ has a nonzero integral). This is a contradiction, and so A has an eigen value.

Topological Proofs

Let $z_0 \in C$ be such that the minimum of $|p(z)|$ on the whole complex plane is achieved at z_0; it was seen at the proof which uses Liouville's theorem that such a number must exist. We can write $p(z)$ as a polynomial in $z - z_0$: there is some natural number k and there are some complex numbers $c_k, c_{k+1}, ..., c_n$ such that $c_k \neq 0$ and that

$$p(z) = p(z_0) + c_k(z - z_0)^k + c_{k+1}(z - z_0)^{k+1} + \cdots + c_n(z - z_0)^n.$$

It follows that if a is a kth root of $-p(z_0)/c_k$ and if t is positive and sufficiently small, then $|p(z_0 + ta)| < |p(z_0)|$, which is impossible, since $|p(z_0)|$ is the minimum of $|p|$ on D.

For another topological proof by contradiction, suppose that $p(z)$ has no zeros. Choose a large positive number R such that, for $|z| = R$, the leading term z^n of $p(z)$ dominates all other terms combined; in other words, such that $|z|^n > |a_{n-1}z^{n-1} + \cdots + a_0|$. As z traverses the circle given by the equation $|z| = R$ once counter-clockwise, $p(z)$, like z^n, winds n times counter-clockwise around 0. At the other extreme, with $|z| = 0$, the "curve" $p(z)$ is simply the single (nonzero) point $p(0)$, whose winding number is clearly 0. If the loop followed by z is continuously deformed between these extremes, the path of $p(z)$ also deforms continuously. We can explicitly write such a deformation as $H(Re^{i\theta},t) = p((1 - t)Re^{i\theta})$ where t is greater than or equal to 0 and less than or equal to 1. If one views the variable t as time, then at time zero the curve is $p(z)$ and at time one the curve is $p(0)$. Clearly at every point t, $p(z)$ cannot be zero by the original assumption, therefore during the deformation, the curve never crosses zero. Therefore the winding number of the curve around zero should never change. However, given that the winding number started as n and ended as 0, this is absurd. Therefore, $p(z)$ has at least one zero.

Algebraic Proofs

These proofs use two facts about real numbers that require only a small amount of analysis (more precisely, the intermediate value theorem):

- every polynomial with odd degree and real coefficients has some real root;
- every non-negative real number has a square root.

The second fact, together with the quadratic formula, implies the theorem for real quadratic polynomials. In other words, algebraic proofs of the fundamental theorem actually show that if R is any real-closed field, then its extension $C = R(\sqrt{-1})$ is algebraically closed.

As mentioned above, it suffices to check the statement "every non-constant polynomial $p(z)$ with real coefficients has a complex root". This statement can be proved by induction on the greatest non-negative integer k such that 2^k divides the degree n of $p(z)$. Let a be the coefficient of z^n in $p(z)$ and let F be a splitting field of $p(z)$ over C; in other words, the field F contains C and there are elements z_1, z_2, ..., z_n in F such that

$$p(z) = a(z - z_1)(z - z_2)\cdots(z - z_n).$$

If $k = 0$, then n is odd, and therefore $p(z)$ has a real root. Now, suppose that $n = 2^k m$ (with m odd and $k > 0$) and that the theorem is already proved when the degree of the polynomial has the form $2^{k-1}m2$ with $m2$ odd. For a real number t, define:

$$q_t(z) = \prod_{1\le i<j\le n} \left(z - z_i - z_j - tz_iz_j\right).$$

Then the coefficients of $q_t(z)$ are symmetric polynomials in the z_i's with real coefficients. Therefore, they can be expressed as polynomials with real coefficients in the elementary symmetric polynomials, that is, in $-a_1$, a_2, ..., $(-1)^n a_n$. So $q_t(z)$ has in fact *real* coefficients. Furthermore, the degree of $q_t(z)$ i n n -1)/2 = $2^{k-1}m(n-1)$, and $m(n-1)$ is an odd number. So, using the induction hypothesis, q_t has at least one complex root; in other words, $z_i + z_j + tz_iz_j$ is complex for two distinct elements i and j from $\{1,...,n\}$. Since there are more real numbers than pairs (i,j), one can find distinct real numbers t and s such that $z_i + z_j + tz_iz_j$ and $z_i + z_j + sz_iz_j$ are complex (for the same i and j). So, both $z_i + z_j$ and z_iz_j are complex numbers. It is easy to check that every complex number has a complex square root, thus every complex polynomial of degree 2 has a complex root by the quadratic formula. It follows that z_i and z_j are complex numbers, since they are roots of the quadratic polynomial $z^2 - (z_i + z_j)z + z_iz_j$.

J. Shipman showed in 2007 that the assumption that odd degree polynomials have roots is stronger than necessary; any field in which polynomials of prime degree have roots is algebraically closed (so "odd" can be replaced by "odd prime" and furthermore this holds for fields of all characteristics). For axiomatisation of algebraically closed fields, this is the best possible, as there are counter examples if a single prime is excluded. However, these counter examples rely on -1 having a square root. If we take a field where "1 has no square root, and every polynomial of degree n - I has a root, where I is any fixed infinite set of odd numbers, then every polynomial $f(x)$ of odd degree has a root (since$(x^2 + 1)^k f(x)$ has a root, where k is chosen so that deg(f) + $2k$ - I).

Another algebraic proof of the fundamental theorem can be given using Galois theory. It suffices to show that C has no proper finite field extension. Let K/C be a finite extension. Since the normal closure of K over R still has a finite degree over C (or R), we may assume without loss of generality that K is a normal extension of R (hence it is a Galois extension, as every algebraic extension of a field of characteristic 0 is separable). Let G be the Galois group of this extension, and let H be a Sylow 2-group of G, so that the order of H is a power of 2, and the index of H in G is odd. By the fundamental theorem of Galois theory, there exists a subextension L of K/R such that Gal(K/L) = H. As [L:R] = [G:H] is odd, and there are no nonlinear irreducible real polynomials of odd degree, we must have L = R, thus [K:R] and [K:C] are powers of 2. Assuming for contradiction [K:C] > 1, the 2-group Gal(K/C) contains a subgroup of index 2, thus there exists a subextension M of C of degree 2. However, C has no extension of degree 2, because every quadratic complex polynomial has a complex root, as mentioned above.

Corollaries

Since the fundamental theorem of algebra can be seen as the statement that the field of complex numbers is algebraically closed, it follows that any theorem concerning algebraically closed fields applies to the field of complex numbers. Here are a few more consequences of the theorem, which are either about the field of real numbers or about the relationship between the field of real numbers and the field of complex numbers:

- The field of complex numbers is the algebraic closure of the field of real numbers.
- Every polynomial in one variable x with real coefficients is the product of a constant, polynomials of the form $x + a$ with a real, and polynomials of the form $x^2 + ax + b$ with a and b real anda^2 - $4b < 0$ (which is the same thing as saying that the polynomial $x^2 + ax + b$ has no real roots).
- Every rational function in one variable x, with real coefficients, can be written as the sum of a polynomial function with rational functions of the form $a/(x - b)^n$ (where n is a natural number, and $a$and b are real numbers), and rational functions of the form $(ax + b)/(x^2 + cx + d)^n$ (where n is a natural number,

and a, b, c, and d are real numbers such that $c^2 - 4d < 0$). A corollary of this is that every rational function in one variable and real coefficients has an elementary primitive.

- Every algebraic extension of the real field is isomorphic either to the real field or to the complex field.

Bounds on The Zeroes of a Polynomial

While the fundamental theorem of algebra states a general existence result, it is of some interest, both from the theoretical and from the practical point of view, to have information on the location of the zeroes of a given polynomial. The simpler result in this direction is a bound on the modulus: all zeroes ζ of a monic polynomial $z^n + a_{n-1}z^{n-1} + \cdots + a_1 z + a_0$ satisfy an inequality $|\zeta| \le R_\infty$. where

$$R_\infty := 1 + \max\{|a_0|, \cdots, |a_{n-1}|\}.$$

Notice that, as stated, this is not yet an existence result but rather an example of what is called an a priori bound: it says that *if there are solutions* then they lie inside the closed disk of centre the origin and radius R_∞. However, once coupled with the fundamental theorem of algebra it says that the disk contains in fact at least one solution. More generally, a bound can be given directly in terms of any p-norm of the n-vector of coefficients $a := (a_0, a_1, \ldots, a_{n-1})$, that is $|\zeta| \le R_p$, where R_p is precisely the q-norm of the 2-vector $(1, \| a \|_p)$, q being the conjugate exponent of p, $1/p + 1/q = 1$, for any $1 \le p \le \infty$. Thus, the modulus of any solution is also bounded by

$$R_1 := \max\left\{1, \sum_{0 \le k < n} |a_k|\right\},$$

$$R_p := \left[1 + \left(\sum_{0 \le k < n} |a_k|^p\right)^{q/p}\right]^{1/q},$$

for $1 < p < \infty$, and in particular

$$R_2 := \sqrt{\sum_{0 \le k \le n} |a_k|^2}$$

(where we define a_n to mean 1, which is reasonable since 1 is indeed the n-th coefficient of our polynomial). The case of a generic polynomial of degree n, $P(z) := a_n z^n + a_{n-1}z^{n-1} + \cdots + a_1 z + a_0$, is of course reduced to the case of a monic, dividing all coefficients by $a_n \ne 0$. Also, in case that 0 is not a root, i.e. $a_0 \neq 0$, bounds from below on the roots ζ follow immediately as bounds from above on $\frac{1}{\zeta}$, that is, the roots of $a_0 z^n + a_1 z^{n-1} + \cdots + a_{n-1} z + a_n$. Finally, the distance $|\zeta - \zeta_0|$ from the roots ζ to any point ζ_0 can be estimated from below and above, seeing $\zeta - \zeta_0$ as zeroes of the polynomial $P(z + \zeta_0)$, whose coefficients are the Taylor expansion of $P(z)$ at $z = \zeta_0$.

We report the here the proof of the above bounds, which is short and elementary. Let ζ be a root of the polynomial $z^n + a_{n-1} z^{n-1} + \cdots + a_1 z + a_0$; in order to prove the inequality $|\zeta| \le R_p$ we can assume, of course, $|\zeta| > 1$. Writing the equation as $-\zeta^n = a_{n-1}\zeta^{n-1} + \cdots + a_1\zeta + a_0$, and using the Hφlder's inequality we find $|\zeta|^n \le \| a \|_p \| (\zeta^{n-1}, \cdots, \zeta, 1) \|_q$. Now, if $p = 1$, this is $|\zeta|^n \le \| a \|_1 \max\{|\zeta|^{n-1}, \cdots, |\zeta|, 1\} = \| a \|_1 |\zeta|^{n-1}$, thus $|\zeta| \le \max\{1, \| a \|_1\}$. In the case $1 < p \le \infty$, taking into account the summation formula for a geometric progression, we have

$$|\zeta|^n \le \| a \|_p \left(|\zeta|^{q(n-1)} + \cdots + |\zeta|^q + 1\right)^{1/q} = \| a \|_p \left(\frac{|\zeta|^{qn} - 1}{|\zeta|^q - 1}\right)^{1/q} \le \| a \|_p \left(\frac{|\zeta|^{qn}}{|\zeta|^q - 1}\right)^{1/q},$$

thus $|\zeta|^{nq} \le \| a \|_p^q \frac{|\zeta|^{qn}}{|\zeta|^q - 1}$ and simplifying, $|\zeta|^q \le 1 + \| a \|_p^q$. Therefore $|\zeta| \le \| (1, \| a \|_p) \|_q = R_p$ holds, for all $1 \le p \le \infty$.

Dirichlet's Unit Theorem

In mathematics, Dirichlet's unit theorem is a basic result in algebraic number theory due to Gustav Lejeune Dirichlet. It determines the rank of the group of units in the ring O_K of algebraic integers of anumber field K. The regulator is a positive real number that determines how "dense" the units are.

Dirichlet's Unit Theorem

The statement is that the group of units is finitely generated and has rank (maximal number of multiplicatively independent elements) equal to

$$r = r_1 + r_2 - 1$$

where r_1 is the *number of real embeddings* and r_2 the *number of conjugate pairs of complex embeddings* of K. This characterisation of r_1 and r_2 is based on the idea that there will be as many ways to embed K in the complex number field as the degree $n = [K : Q]$; these will either be into the real numbers, or pairs of embeddings related by complex conjugation, so that

$$n = r_1 + 2r_2.$$

Other ways of determining r_1 and r_2 are

- use the primitive element theorem to write $K = Q(\alpha)$, and then r_1 is the number of conjugates of α that are real, $2r_2$ the number that are complex;
- write the tensor product of fields $K \Delta_Q R$ as a product of fields, there being r_1 copies of R and *r2 copies* of C.

As an example, if K is a quadratic field, the rank is 1 if it is a real quadratic field, and 0 if an imaginary quadratic field. The theory for real quadratic fields is essentially the theory of Pell's equation.

The rank is > 0 for all number fields besides Q and imaginary quadratic fields, which have rank 0. The 'size' of the units is measured in general by a determinant called the regulator. In principle a basis for the units can be effectively computed; in practice the calculations are quite involved when n is large.

The torsion in the group of units is the set of all roots of unity of K, which form a finite cyclic group. For a number field with at least one real embedding the torsion must therefore be only {1,-1}. There are number fields, for example most imaginary quadratic fields, having no real embeddings which also have {1,-1} for the torsion of its unit group. Totally real fields are special with respect to units. If L/K is a finite extension of number fields with degree greater than 1 and the units groups for the integers of L and K have the same rank then K is totally real and L is a totally complex quadratic extension. The converse holds too. (An example is Kequal to the rationals and L equal to an imaginary quadratic field; both have unit rank 0.)

There is a generalisation of the unit theorem by Helmut Hasse (and later Claude Chevalley) to describe the structure of the group of *S-units*, determining the rank of the unit group in localisations of rings of integers. Also, the Galois module structure of $Q \oplus O_{K,S} \otimes_Z Q$ has been determined.

The Regulator

Suppose that $u_1, \ldots, u_r$ are a set of generators for the unit group modulo roots of unity. If u is an algebraic number, write $u^1, \ldots, u^{r+1}$ for the different embeddings into R or C, and write N_i for the degree of the corresponding embedding over R (so it is 1 for real embeddings and 2 for complex ones). Then the r by $r + 1$ matrix whose entries are $N_i \log | u_i^j |$ has the property that the sum of any row is zero (because all units have norm 1, and the log of the norm is the sum of the entries of a row). This implies that the absolute value R of the determinant of the submatrix formed by deleting one column is independent of the column. The number R is called the regulator of the algebraic number field (it does not depend on the choice of generators u_i). It measures the "density" of the units: if the regulator is small, this means that there are "lots" of units.

The regulator has the following geometric interpretation. The map taking a unit u to the vector with entries $N_i \log | u_i |$ has image in the r-dimensional subspace of R^{r+1} consisting of all vectors whose entries have sum 0, and by Dirichlet's unit theorem the image is a lattice in this subspace. The volume of a fundamental domain of this lattice is $R\Phi(r+1)$.

The regulator of an algebraic number field of degree greater than 2 is usually quite cumbersome to calculate, though there are now computer algebra packages that can do it in many cases. It is usually much easier to calculate the product hR of the class number h and the regulator using the class number formula, and the main difficulty in calculating the class number of an algebraic number field is usually the calculation of the regulator.

Examples

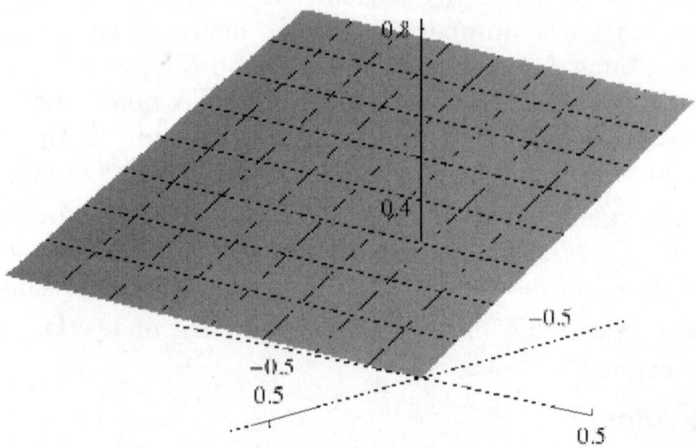

Figure 1: A fundamental domain in logarithmic space of the group of units of the cyclic cubic field K obtained by adjoining to Q a root of $f(x) = x^3 + x^2 - 2x - 1$. If α denotes a root of $f(x)$, then a set of fundamental units is $\{\varepsilon_1, \varepsilon_2\}$ where $\varepsilon_1 = \alpha^2 + \alpha - 1$ and $\varepsilon_2 = 2 - \alpha^2$. The area of the fundamental domain is approximately 0.910114, so the regulator of K is approximately 0.525455.

- The regulator of an imaginary quadratic field, or of the rational integers, is 1 (as the determinant of a 0×0 matrix is 1).
- The regulator of the real quadratic field Q(ϑ5) is log((ϑ5 + 1)/ 2). This can be seen as follows. A fundamental unit is (ϑ5 + 1)/2, and its images under the two embeddings into R are (ϑ5 + 1)/2 and (-ϑ5 + 1)/2. So the *r* by *r* + 1 matrix is

$$\left[1\times\log\left|\frac{\sqrt{5}+1}{2}\right|,\ 1\times\log\left|\frac{-\sqrt{5}+1}{2}\right|\right].$$

- The regulator of the cyclic cubic field Q(α), where α is a root of $x^3 + x^2 - 2x - 1$, is approximately 0.5255. A basis of the group of units modulo roots of unity is $\{\varepsilon_1, \varepsilon_2\}$ where $\varepsilon_1 = \alpha^2 + \alpha - 1$ and $\varepsilon_2 = 2 - \alpha^2$.

Higher Regulators

A 'higher' regulator refers to a construction for an algebraic K-group with index $n > 1$ that plays the same role as the classical regulator does for the group of units, which is a group K_1. A theory of such regulators has been in development, with work of Armand Borel and others. Such higher regulators play a role, for example, in the Beilinson conjectures, and are expected to occur in evaluations of certain L-functions at integer values of the argument.

Stark Regulator

The formulation of Stark's conjectures led Harold Stark to define what is now called the Stark regulator, similar to the classical regulator as a determinant of logarithms of units, attached to any Artin representation.

Artin Reciprocity Law

The Artin reciprocity law, established by Emil Artin in a series of papers (1924; 1927; 1930), is a general theorem in number theory that forms a central part of the global class field theory. The term "reciprocity law" refers to a long line of more concrete number theoretic statements which it generalised, from the quadratic reciprocity law and the reciprocity laws of Eisenstein and Kummer toHilbert's product formula for the norm symbol. Artin's result provided a partial solution to Hilbert's ninth problem.

Significance

Artin's reciprocity law implies a description of the abelianisation of the absolute Galois group of aglobal field K *which is based on the local–global principle and the use of the Frobenius elements. Together with the Takagi existence theorem, it is used to describe the abelian extensions of* K *in terms of arithmetic of* K *and to understand the behaviour of the nonarchimedean places in them. Therefore, the Artin reciprocity law can be interpreted as one of the main theorems of the global class field theory. It can be used to prove that Artin L-functions are meromorphic and for the proof of the Chebotarev density theorem.*

Finite Extensions of Global Fields

The definition of the Artin map for a finite (abelian) extension L/K *of global fields (such as a finite abelian extension of Q) has a concrete description in terms of prime ideals and Frobenius elements.*

If $\mathfrak{p}$ is a prime of K then the decomposition groups of primes $\mathfrak{P}$ above $\mathfrak{p}$ are equal in Gal(L/K) since the latter group is abelian. If $\mathfrak{p}$ is unramified in L, then the decomposition group $D_\mathfrak{p}$ is canonically isomorphic to the Galois group of the extension of residue fields $\mathcal{O}_{L,\mathfrak{P}} / \mathfrak{P}$ over $\mathcal{O}_{K,\mathfrak{p}} / \mathfrak{p}$. There is therefore a canonically defined Frobenius element in Gal(L/K) denoted by $\mathrm{Frob}_\mathfrak{p}$ or $\left(\frac{\mathfrak{p}}{L/K}\right)$. If Δ denotes the relative discriminant of L/K, the Artin symbol (or Artin map, or (global) reciprocity map) of L/K is defined on the group of prime-to-Δ fractional ideals, I_K^Δ, by linearity:

$$\left(\frac{\cdot}{L/K}\right): \quad I_K^\Delta \quad \to \quad \mathrm{Gal}(L/K)$$

$$\prod_{i=1}^{m} \mathfrak{p}_i^{n_i} \quad \mapsto \quad \prod_{i=1}^{m} \left(\frac{\mathfrak{p}_i}{L/K}\right)^{n_i}.$$

The Artin reciprocity law (or global reciprocity law) states that there is a modulus c of K such that the Artin map induces an isomorphism

$$I_K^c / i(K_{c,1}) \mathrm{Nm}_{L/K}(I_L^c) \xrightarrow{\sim} \mathrm{Gal}(L/K)$$

where $K_{c,1}$ is the ray modulo c, $\mathrm{Nm}_{L/K}$ is the norm map associated to L/K and I_L^c is the fractional ideals of L prime to c. Such a modulus c is called a defining modulus for L/K. The smallest defining modulus is called the conductor of L/K and typically denoted $\mathfrak{f}(L/K)$.

Example: Quadratic Fields

If d is a squarefree integer, $K = \mathbb{Q}$, and $L = \mathbb{Q}(\sqrt{d})$, then the Galois group Gal($L/\mathbb{Q}$) can be identified with $\{\pm 1\}$. The discriminant of L over $\mathbb{Q}$ is d or $4d$ depending on whether $d \equiv 1 \pmod 4$ or not. The Artin map is then defined on primes p that do not divide Δ by

$$p \mapsto \left(\frac{d}{p}\right)$$

where $\left(\frac{d}{p}\right)$ is the Legendre symbol.

Cohomological Interpretation

Let $L_v \supset K_v$ *be a Galois extension of local fields with Galois group* G. *The local reciprocity law describes a canonical isomorphism*

$$\theta_v : K_v^\times / N_{L_v/K_v}(L_v^\times) \to G^{ab},$$

called the local Artin symbol.

Let $L \supset K$ be a Galois extension of global fields and C_L stand for the idèle class group of L. The maps θ_v fo r different places v of K can be assembled into a single global symbol map by multiplying the local components of an idèle class. One of the statements of the Artin reciprocity law is that this results in the canonical isomorphism

$$\theta : C_K / N_{L/K}(C_L) \to \mathrm{Gal}(L/K)^{ab}.$$

A cohomological proof of the global reciprocity law can be achieved by first establishing that

$$(\mathrm{Gal}(K^{sep}/K), \varinjlim C_L)$$

constitutes a class formation in the sense of Artin and Tate. Then one proves that

$$\hat{H}^0(\mathrm{Gal}(L/K), C_L) \simeq \hat{H}^{-2}(\mathrm{Gal}(L/K), \mathbb{Z}),$$

where $\hat{H}^i$ denote the Tate cohomology groups. Working out the cohomology groups establishes that θ is an isomorphism.

Alternative Statement

An alternative version of the reciprocity law, leading to the Langlands program, connects Artin L-functions associated to abelian extensions of a number field with Hecke L-functions associated to characters of the idèle class group.

A Hecke character (or Grφßencharacter) of a number field *K* is defined to be a quasicharacter of the idèle class group of *K*. Robert Langlands interpreted Hecke characters as automorphic forms on the reductive algebraic group *GL*(1) over the ring of adeles of *K*.

Let *E*D *K* be an abelian Galois extension with Galois group *G*. Then for any character σ: *G* → $C^{\times}$ (i.e. one-dimensional complex representation of the group *G*), there exists a Hecke character χ of *K* such that

$$L_{E/K}^{Artin}(\sigma,s) = L_{K}^{Hecke}(\chi,s)$$

where the left hand side is the Artin L-function associated to the extension with character σ and the right hand side is the Hecke L-function associated with χ, Section 7.D of.

The formulation of the Artin reciprocity law as an equality of *L*-functions allows formulation of a generalisation to *n*-dimensional representation, though a direct correspondence still lacking.

Chapter 5

Important Theorems in Algebraic Topology

Borsuk–Ulam Theorem

In mathematics, the Borsuk–Ulam theorem, named after Stanisaw Ulam and Karol Borsuk, states that every continuous function from an n-sphere into Euclidean n-space maps some pair of antipodal points to the same point. Here, two points on a sphere are called antipodal if they are in exactly opposite directions from the sphere's centre. According to (Matoušek 2003), the first historical mention of the statement of this theorem appears in (Lyusternik 1930).

The first proof was given by (Borsuk 1933), where the formulation of the problem was attributed to Ulam. Since then, many alternate proofs have been found out by various authors as collected in.

The case $n = 2$ is often illustrated by saying that at any moment there is always a pair of antipodal points on the Earth's surface with equal temperatures and equal barometric pressures. This assumes that temperature and barometric pressure vary continuously.

A stronger statement related to Borsuk–Ulam theorem is that every antipode-preserving map f from S^n to itself has odd degree.

Corollaries

- No subset of R^n is homeomorphic to S^n.
- The Lusternik–Schnirelmann theorem: If the sphere S^n is covered by $n + 1$ open sets, then one of these sets contains a pair (x, "x) of antipodal points. (this is equivalent to the Borsuk–Ulam theorem)
- The Ham sandwich theorem: For any compact sets $A_1,\ldots,A_n$ in R^n we can always find a hyperplane dividing each of them into two subsets of equal measure).
- The Brouwer fixed point theorem (Matoušek 2003; Su 1997).

Brouwer Fixed-point Theorem

Brouwer's fixed-point theorem is a fixed-point theoremin topology, named after Luitzen Brouwer. It states that for any continuous function f with certain properties there is a point x_0 such that $f(x_0) = x_0$. The simplest form of Brouwer's theorem is for continuous functions f from a disk D to itself. A more general form is for continuous functions from aconvex compact subset K of Euclidean space to itself.

Among hundreds of fixed-point theorems, Brouwer's is particularly well known, due in part to its use across numerous fields of mathematics. In its original field, this result is one of the key theorems characterising the topology of Euclidean spaces, along with the Jordan curve theorem, the hairy ball theorem and the Borsuk–Ulam theorem. This gives it a place among the fundamental theorems of topology. The theorem is also used for proving deep results about differential equations and is covered in most introductory courses on differential geometry. It appears in unlikely fields such as game theory. In economics, Brouwer's fixed-point theorem and its extension, the Kakutani fixed-point theorem, play a central role in the proof of existence of general equilibrium in market economies as developed in the 1950s by economics Nobel prize winners Gerard Debreu andKenneth Arrow.

The theorem was first studied in view of work on differential equations by the French mathematicians around Poincaré and Picard. Proving results such as the Poincaré–Bendixson theorem requires the use of topological methods. This work at the end of the 19th century opened into several successive versions of the theorem. The general case was first proved in 1910 by Hadamard, and then in 1912 by Luitzen Egbertus Jan Brouwer.

Statement

The theorem has several formulations, depending on the context in which it is used. The simplest is sometimes given as follows:

In the Plane: Every continuous function f from a closed disk to itself has at least one fixed point.

This can be generalised to an arbitrary finite dimension:

In Euclidean space: Every continuous function from a closed ball of a Euclidean space to itself has a fixed point.

A slightly more general version is as follows:

Convex Compact Set: Every continuous function f from a convex compact subset K of a Euclidean space to K itself has a fixed point.

An even more general form is better known under a different name:

Schauder Fixed Point Theorem: Every continuous function from a convex compact subset K of a Banach space to K itself has a fixed point.

Notes

The function f in this theorem is not required to be bijective or even surjective. Since any closed ball in Euclidean n-space is homeomorphic to the closed unit ball D^n, the theorem also has equivalent formulations that only state it for D^n.

Because the properties involved (continuity, being a fixed point) are invariant under homeomorphisms, the theorem is equivalent to forms in which the domain is required to be a closed unit ball D^n. For the same reason it holds for every set that is homeomorphic to a closed ball (and therefore also closed, bounded, connected, without holes, etc.).

The statement of the theorem is false if formulated for the *open* unit disk, the set of points with distance strictly less than 1 from the origin. Consider for example the function

$$f(x) = (x + 1) / 2$$

which is a continuous function from the open interval (-1,1), the one-dimensional unit disk, to itself. As it shifts every point to the right, it cannot have a fixed point. (But it does have a fixed point for the closed interval [-1,1], namely f(x) = x = 1).

Illustrations

The theorem has several "real world" illustrations. For example: take two sheets of graph paper of equal size with coordinate systems on them, lay one flat on the table and crumple up (without ripping or tearing) the other one and place it, in any fashion, on top of the first so that the crumpled paper does not reach outside the flat one. There will then be at least one point of the crumpled sheet that lies directly above its corresponding point (i.e. the point with the same coordinates) of the flat sheet. This is a consequence of the n = 2 case of Brouwer's theorem applied to the continuous map that assigns to the coordinates of every point of the crumpled sheet the coordinates of the point of the flat sheet immediately beneath it.

Similarly: Take an ordinary map of a country, and suppose that that map is laid out on a table inside that country. There will always be a "You are Here" point on the map which represents that same point in the country. In three dimensions the consequence of the Brouwer fixed-point theorem is that no matter how much you stir a cocktail in a glass some point in the liquid will remain in exactly the same place in the glass as before you took any action, assuming that, presumably, the final position of each point is a continuous function of its original position, and that the liquid after stirring is contained within the space originally taken up by it.

Intuitive approach

Explanations Attributed to Brouwer: The theorem is supposed to have originated from Brouwer's observation of a cup of coffee. If one stirs to dissolve a lump of sugar, it appears there is always a point without motion. He drew the conclusion that at any moment, there is a point on the surface that is not moving. The fixed point is not necessarily the point that seems to be motionless, since the centre of the turbulence moves a little bit. The result is not intuitive, since the original fixed point may become mobile when another fixed point appears.

Brouwer is said to have added: "I can formulate this splendid result different, I take a horizontal sheet, and another identical one which I crumple, flatten and place on the other. Then a point of the crumpled sheet is in the same place as on the other sheet." Brouwer "flattens" his sheet as with a flat iron, without removing the folds and wrinkles.

One-dimensional Case

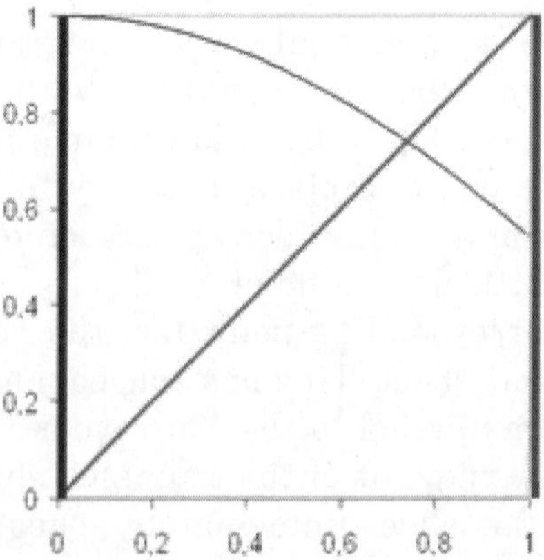

In one dimension, the result is intuitive and easy to prove. The continuous function f is defined on a closed interval $[a, b]$ and takes values in the same interval. Saying that this function has a fixed point amounts to saying that its graph intersects that of the function defined on the same interval $[a, b]$ which maps x to x (green).

Intuitively, any continuous line from the left edge of the square to the right edge must necessarily intersect the green diagonal. It is not hard to give a formal proof. It suffices to consider the function g which maps x to $f(x) - x$. It is ≥ 0 on a and ≤ 0 on b. By the intermediate value theorem, g has a zero in $[a, b]$; this zero is a fixed point.

Brouwer is said to have expressed this as follows: "Instead of examining a surface, we will prove the theorem about a piece of string. Let us begin with the string in an unfolded state, then refold it. Let us flatten the refolded string. Again a point of the string has not changed its position with respect to its original position on the unfolded string." In one dimension, Brouwer's fixed point theorem is equivalent to the intermediate value theorem.

History

The Brouwer fixed point theorem was one of the early achievements of algebraic topology, and is the basis of more general fixed point theorems which are important in functional analysis. The case n = 3 first was proved by Piers Bohl in 1904 (published in Journal für die reine und angewandte Mathematik). It was later proved by L. E. J. Brouwer in 1909. Jacques Hadamard proved the general case in 1910, and Brouwer found a different proof in 1912. Since these early proofs were all non-constructive indirect proofs, they ran contrary to Brouwer's intuitionist ideals. Methods to construct (approximations to) fixed points guaranteed by Brouwer's theorem are now known, however;

Prehistory

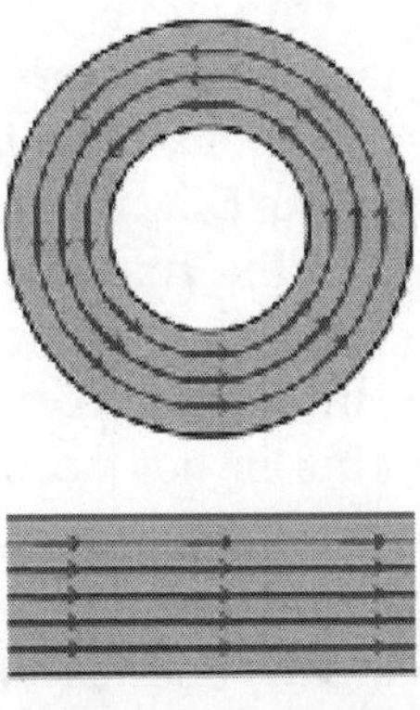

Figure 1: For flows in an unbounded area, or in an area with a "hole", the theorem is not applicable.

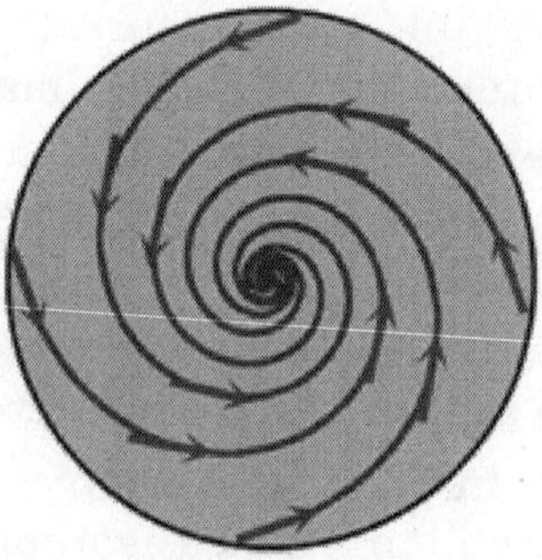

Figure 2: The theorem applies to any disk-shaped area, where it guarantees the existence of a fixed point.

To understand the prehistory of Brouwer's fixed point theorem one needs to pass through differential equations. At the end of the 19th century, the old problem of the stability of the solar system returned into the focus of the mathematical community. Its solution required new methods. As noted by Henri Poincaré, who worked on the three-body problem, there is no hope to find an exact solution: "Nothing is more proper to give us an idea of the hardness of the three-body problem, and generally of all problems of Dynamics where there is no uniform integral and the Bohlin series diverge." He also noted that the search for an approximate solution is no more efficient: "the more we seek to obtain precise approximations, the more the result will diverge towards an increasing imprecision.".

He studied a question analogous to that of the surface movement in cup of coffee. What can we say, in general, about the trajectories on a surface animated by a constant flow? Poincaré discovered that the answer can be found in what we now call the topological properties in the area containing the trajectory. If this area is compact, i.e. both closed and bounded, then the trajectory either becomes stationary, or it approaches a limit cycle. Poincaré went further; if the area is of the same kind as a disk, as is the case for the cup of coffee, there must necessarily be a fixed point. This fixed point is invariant under all functions which associate to each point of the original surface its position after a short time interval t. If the area is a circular band, or if it is not closed, then this is not necessarily the case.

To understand differential equations better, a new branch of mathematics was born. Poincaré called it analysis *situs*. The French Encyclopaedia Universalis defines it as the branch which "treats the properties of an object that are invariant if it is deformed in any continuous way, without tearing". In 1886, Poincaré proved a result that is equivalent to Brouwer's fixed-point theorem, although thc connection with the subject of this article was not yet apparent. A little later, he developed one of the fundamental tools for better understanding the analysis situs, now known as the fundamental group or sometimes the Poincaré group. This method can be used for a very compact proof of the theorem under discussion.

Poincaré's method was analogous to that of Émile Picard, a contemporary mathematician who generalised the Cauchy–Lipschitz theorem. Picard's approach is based on a result that would later be formalised by another fixed-point theorem, named after Banach. Instead of the topological properties of the domain, this theorem uses the fact that the function in question is a contraction.

First Proofs

At the dawn of the 20th century, the interest in analysis situs did not stay unnoticed. However, the necessity of a theorem equivalent to the one discussed in this article was not yet evident. Piers Bohl, a Latvian mathematician, applied topological methods to the study of differential equations. In 1904 he proved the three-dimensional case of our theorem, but his publication was not noticed.

It was Brouwer, finally, who gave the theorem its first patent of nobility. His goals were different from those of Poincaré. This mathematician was inspired by the foundations of mathematics, especially mathematical logic and topology. His initial interest lay in an attempt to solve Hilbert's fifth problem. In 1909, during a voyage to Paris, he met Poincaré, Hadamard and Borel.

The ensuing discussions convinced Brouwer of the importance of a better understanding of Euclidean spaces, and were the origin of a fruitful exchange of letters with Hadamard. For the next four years, he concentrated on the proof of certain great theorems on this question. In 1912 he proved the hairy ball theorem for the two-dimensional sphere, as well as the fact that every continuous map from the two-dimensional ball to itself has a fixed point. These two results in themselves were not really new. As Hadamard observed, Poincaré had shown a theorem equivalent to the hairy ball theorem.

The revolutionary aspect of Brouwer's approach was his systematic use of recently developed tools such as homotopy, the underlying concept of the Poincare group. In the following year, Hadamard generalised the theorem under discussion to an arbitrary finite dimension, but he employed different methods. H. Freudenthal comments on the respective roles as follows: "Compared to Brouwer's revolutionary methods, those of Hadamard were very traditional, but Hadamard's participation in the birth of Brouwer's ideas resembles that of a midwife more than that of a mere spectator.".

Brouwer's approach yielded its fruits, and in 1912 he also found a proof that was valid for any finite dimension., as well as other key theorems such as the invariance of dimension. In the context of this work, Brouwer also generalised the Jordan curve theorem to arbitrary dimension and established the properties connected with the degree of a continuous mapping. This branch of mathematics, originally envisioned by Poincaré and developed by Brouwer, changed its name. In the 1930s, analysis situs became algebraic topology.

Brouwer's celebrity is not exclusively due to his topological work. He was also the originator and zealous defender of a way of formalising mathematics that is known as intuitionism, which at the time made a stand against set theory. While Brouwer preferred constructive proofs, ironically, the original proofs of his great topological theorems were not constructive, and it took until 1967 for constructive proofs to be found.

Reception

The theorem proved its worth in more than one way. During the 20th century numerous fixed-point theorems were developed, and even a branch of mathematics called fixed-point theory. Brouwer's theorem is probably the most important. It is also among the foundational theorems on the topology of topological manifolds and is often used to prove other important results such as the Jordan curve theorem.

Besides the fixed-point theorems for more or lesscontracting functions, there are many that have emerged directly or indirectly from the result under discussion. A continuous map from a closed ball of Euclidean space to its boundary cannot be the identity on the boundary. Similarly, the Borsuk–Ulam theorem says that a continuous map from the n-dimensional sphere to R^n has a pair of antipodal points that are mapped to the same point. In the finite-dimensional case, the Lefschetz fixed-point theorem provided from 1926 a method for counting fixed points. In 1930, Brouwer's fixed-point theorem was generalised to Banach spaces. This generalisation is known as Schauder's fixed-point theorem, a result generalised further by S. Kakutani to multivalued functions. One also meets the theorem and its variants outside topology. It can be used to prove the Hartman-Grobman theorem, which describes the qualitative behaviour of certain differential equations near certain equilibria. Similarly, Brouwer's theorem is used for the proof of the théorème de la variété centrale. The theorem can also be found in existence proofs for the solutions of certain partial differential equations.

Other areas are also touched. In game theory, John Nash used the theorem to prove that in the game of Hex there is a winning strategy for white. In economy, P. Bich explains that certain generalisations of the theorem show that its use is helpful for certain classical problems in game theory and generally for equilibria (Hotelling's law), financial equilibria and incomplete markets.

Proof Outlines

A Proof Using Homology

The proof uses the observation that the boundary of D^n is S^{n-1}, the $(n - 1)$-sphere.

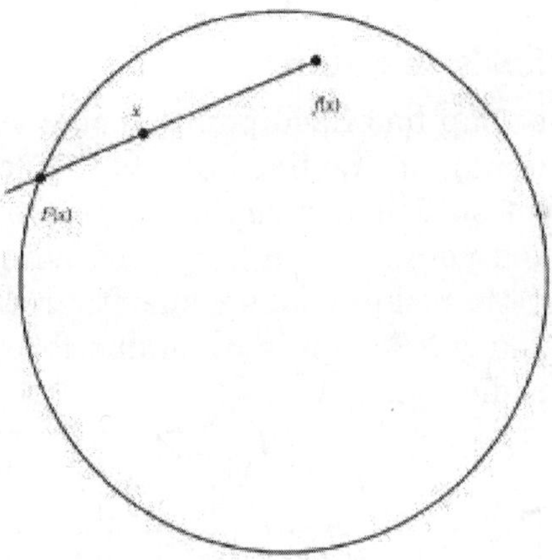

Figure 3: Illustration of the retraction F

The argument proceeds by contradiction, supposing that a continuous function $f: D^n \to D^n$ has *no* fixed point, and then attempting to derive an inconsistency, which proves that the function must in fact have a fixed point. For each $x$in D^n, there is only one straight line that passes through$f(x)$ and x, because it must be the case that $f(x)$ and x are distinct by hypothesis (recall that f having no fixed points means that $f(x) \neq x$). Following this line from $f(x)$ through $x$leads to a point on S^{n-1}, denoted by $F(x)$. This defines a continuous function $F: D^n \to S^{n-1}$, which is a special type of continuous function known as a retraction:

every point of the codomain (in this case S^{n-1}) is a fixed point of the function.

Intuitively it seems unlikely that there could be a retraction of D^n onto S^{n-1}, and in the case $n = 1$ it is obviously impossible because S^0 (i.e., the endpoints of the closed interval D^1) is not even connected. The case$n = 2$ is less obvious, but can be proven by using basic arguments involving the fundamental groups of the respective spaces: the retraction would induce an injective group homomorphism from the fundamental group of S^1 to that of D^2, but the first group is isomorphic to Z while the latter group is trivial, so this is impossible. The case $n = 2$ can also be proven by contradiction based on a theorem about non-vanishing vector fields.

For $n > 2$, however, proving the impossibility of the retraction is more difficult. One way is to make use of homology groups: the homology $H_{n-1}(D^n)$ is trivial, while $H_{n-1}(S^{n-1})$ is infinite cyclic. This shows that the retraction is impossible, because again the retraction would induce an injective group homomorphism from the latter to the former group.

A Proof Using Stokes's Theorem

To prove that a map has fixed points, one can assume that it is smooth, because if a map has no fixed points then convoluting it with a smooth function of sufficiently small support produced a smooth function with no fixed points. As in the proof using homology, one is reduced to proving that there is no smooth retraction f from the ball B onto its boundary dB. If ω is a volume form on the boundary then giving a contradiction.

$0 <$	$\int$	$\omega =$	$\int$	$f^*(\omega) =$	$\int$	$df^*(\omega) =$	$\int$	$f^*(d\omega) =$	$\int$	$f^*(0) = 0$
	dB		dB		B		B		B	

More generally, this shows that there is no smooth retraction from any non-empty smooth orientable compact manifold onto its boundary. The proof using Stokes's theorem is closely related to the proof using homology (or rather cohomology), because the form ω generates the de Rham cohomology group $H^{n-1}(dB)$ used in the cohomology proof.

A Combinatorial Proof

There is also a more elementary combinatorial proof, whose main step consists in establishing Sperner's lemma in n dimensions.

A Proof by Hirsch

There is also a quick proof, by Morris Hirsch, based on the impossibility of a differentiable retraction. The indirect proof starts by noting that the map *f* can be approximated by a smooth map retaining the property of not fixing a point; this can be done by using the Weierstrass approximation theorem, for example. One then defines a retraction as above which must now be differentiable. Such a retraction must have a non-singular value, by Sard's theorem, which is also non-singular for the restriction to the boundary (which is just the identity). Thus the inverse image would be a 1-manifold with boundary. The boundary would have to contain at least two end points, both of which would have to lie on the boundary of the original ball—which is impossible in a retraction.

Kellogg, Li, and Yorke turned Hirsch's proof into a constructive proof by observing that the retract is in fact defined everywhere except at the fixed points. For almost any point, q, on the boundary, (assuming it is not a fixed point) the one manifold with boundary mentioned above does exist and the only possibility is that it leads from q to a fixed point. It is an easy numerical task to follow such a path from q to the fixed point so the method is essentially constructive. Chow, Mallet-Paret, and Yorke gave a conceptually similar path-following version of the homotopy proof which extends to a wide variety of related problems.

A Proof Using the Game Hex

A quite different proof given by David Gale is based on the game of Hex. The basic theorem about Hex is that no game can end in a draw. This is equivalent to the Brouwer fixed-point theorem for dimension 2. By considering n-dimensional versions of Hex, one can prove in general that Brouwer's theorem is equivalent to the determinacy theorem for Hex.

A Proof Using the Lefschetz Fixed-point Theorem

The Lefschetz fixed-point theorem says that if a continuous map *f* from a finite simplicial complex *B* to itself has only isolated fixed points, then the number of fixed points counted with multiplicities (which may be negative) is equal to the Lefschetz number

$$\sum_{n}(-1)^{n}\mathrm{Tr}(f \mid H_{n}(B))$$

and in particular if the Lefschetz number is nonzero then *f* must have a fixed point. If *B* is a ball (or more generally is contractible) then the

Lefschetz number is one because the only non-zero homology group is $H^0(B)$, so f has a fixed point.

A Proof in a Weak Logical System

In reverse mathematics, Brouwer's theorem can be proved in the system WKL_0, and conversely over the base system RCA_0 Brouwer's theorem for a square implies the weak Konig's lemma, so this gives a precise description of the strength of Brouwer's theorem.

Generalisations

The Brouwer fixed-point theorem forms the starting point of a number of more general fixed-point theorems.

The straightforward generalisation to infinite dimensions, i.e. using the unit ball of an arbitrary Hilbert space instead of Euclidean space, is not true. The main problem here is that the unit balls of infinite-dimensional Hilbert spaces are not compact. For example, in the Hilbert space ℓ^2 of square-summable real (or complex) sequences, consider the map $f: \ell^2 \rightarrow \ell^2$ which sends a sequence (x_n) from the closed unit ball of ℓ^2 to the sequence (y_n) defined by

$$y_0 = \sqrt{1-||x||_2^2} \qquad \text{and} \qquad y_n = x_{n-1} \quad \text{for} \quad n \geq 1.$$

It is not difficult to check that this map is continuous, has its image in the unit sphere of ℓ^2, but does not have a fixed point.

The generalisations of the Brouwer fixed-point theorem to infinite dimensional spaces therefore all include a compactness assumption of some sort, and in addition also often an assumption of convexity.

There is also finite-dimensional generalisation to a larger class of spaces: If X is a product of finitely many chainable continua, then every continuous function $f: X \rightarrow X$ *has a fixed point, where a chainable continuum is a (usually but in this case not necessarily metric) compact Hausdorff space of which every open cover has a finite open refinement* $\{U_1,\ldots,U_m\}$, *such that* $U_i \cap U_j \neq \varnothing$ *if and only if* $|i-j| \leq 1$. *Examples of chainable continua include compact connected linearly ordered spaces and in particular closed intervals of real numbers.*

The Kakutani fixed point theorem generalises the Brouwer fixed-point theorem in a different direction: it stays in R^n, but considers upper semi-continuous correspondences (functions that assign to each point of the set a subset of the set). It also requires compactness and convexity of the set.

The Lefschetz fixed-point theorem applies to (almost) arbitrary compact topological spaces, and gives a condition in terms of singular homology that guarantees the existence of fixed points; this condition is trivially satisfied for any map in the case of D^n.

Cellular Approximation

In algebraic topology, in the cellular approximation theorem, a map between CW-complexes can always be taken to be of a specific type. Concretely, if X and Y are CW-complexes, and $f : X \to Y$ is a continuous map, then f is said to be *cellular*, if f takes the n-skeleton of X to the n-skeleton of Y for all n, i.e. if $f(X^n) \subseteq Y^n$ for all n. The content of the cellular approximation theorem is then that any continuous map $f : X \to Y$ between CW-complexes X and Y is homotopic to a cellular map, and if f is already cellular on a subcomplex A of X, then we can furthermore choose the homotopy to be stationary on A. From an algebraic topological viewpoint, any map between CW-complexes can thus be taken to be cellular.

Idea of Proof

The proof can be given by induction after n, *with the statement that* f *is cellular on the skeleton* X^n. *For the base case n=0, notice that every path-component of* Y *must contain a 0-cell. The image under* f *of a 0-cell of* X *can thus be connected to a 0-cell of* Y *by a path, but this gives a homotopy from* f *to a map, which is cellular on the 0-skeleton of X.*

Assume inductively that f *is cellular on the* (n - 1)*-skeleton of* X, *and let* e^n *be an* n*-cell of* X. *Theclosure of* e^n *is compact in* X, *being the image of the characteristic map of the cell, and hence the image of the closure of* e^n *under* f *is also compact in* Y. *Then it is a general result of CW-complexes that any compact subspace of a CW-complex meets (that is, intersects non-trivially) only finitely many cells of the complex. Thus* $f(e^n)$ *meets at most finitely many cells of* Y, *so we can take* $e^k \subseteq Y$ *to be a cell of highest dimension meeting* $f(e^n)$. *If* $k \leq n$, *the map* f *is already cellular on* e^n, *since in this case only cells of the* n*-skeleton of* Y *meets* $f(e^n)$, *so we may assume that* $k > n$. *It is then a technical, non-trivial result that the restriction of* f *to* $X^{n-1} \cup e^n$ *can be homotoped relative to* X^{n-1} *to a map missing a point* p-e^k. *Since* Y^k -p} *deformation retracts onto the subspace* Y^k-e^k, *we can further homotope the restriction of* f *to* $X^{n-1} \cup e^n$ *to a map, say,* g, *with the property that* $g(e^n)$ *misses the* *cell* e^k *of* Y, *still relative to* X^{n-1}. *Since* $f(e^n)$ *met only finitely many cells of* Y *to begin with, we can repeat this process finitely many times to make* $f(e^n)$ *miss all cells of* Y *of dimension larger than* n.

We repeat this process for every *n*-cell of *X*, fixing cells of the subcomplex *A* on which *f* is already cellular, and we thus obtain a homotopy (relative to the (*n* - 1)-skeleton of *X* and the *n*-cells of *A*) of the restriction of *f* to X^n to a map cellular on all cells of *X* of dimension at most *n*. Using then the homotopy extension property to extend this to a homotopy on all of *X*, and patching these homotopies together, will finish the proof. For details, consult Hatcher.

Applications

Some Homotopy Groups

The cellular approximation theorem can be used to immediately calculate some homotopy groups. In particular, if $n < k$, then $\pi_n(S^k) = 0$: Give S^n and S^k their canonical CW-structure, with one 0-cell each, and with one *n*-cell for S^n and one *k*-cell for S^k. Any base-point preserving map *f*: $S^n \to S^k$ is by the cellular approximation theorem homotopic to a constant map, whence $\pi_n(S^k) = 0$.

Cellular Approximation for Pairs

Let $f:(X,A) \to (Y,B)$ be a map of CW-pairs, that is, *f* is a map from *X* to *Y*, and the image of $A \subseteq X$ under *f* sits inside *B*. Then *f* is homotopic to a cellular map $(X,A) \to (Y,B)$. The restrict *f* to *A* and use cellular approximation to obtain a homotopy of *f* to a cellular map on *A*. Use the homotopy extension property to extend this homotopy to all of *X*, and apply cellular approximation again to obtain a map cellular on *X*, but without violating the cellular property on *A*.

As a consequence, we have that a CW-pair *(X,A)* is n-connected, if all cells of *X - A* have dimension strictly greater than *n*: If $i \le n$, then any map $(D^i, \partial D^i) \to (X,A)$ is homotopic to a cellular map of pairs, and since the *n*-skeleton of *X* sits inside *A*, any such map is homotopic to a map whose image is in *A*, and hence it is 0 in the relative homotopy group $\pi_i(X,A)$.

We have in particular that (X,X^n) is n-connected, so it follows from the long exact sequence of homotopy groups for the pair (X,X^n) that we have isomorphisms $\pi_i(X^n) \to \pi_i(X)$ for all $i < n$ and a surjection $\pi_n(X^n) \to \pi_n(X)$.

Eilenberg–Zilber Theorem

In mathematics, specifically in algebraic topology, the Eilenberg–Zilber theorem is an important result in establishing the link between the homology groups of a product space $X \times Y$ and those of the spaces X and Y. The theorem first appeared in a 1953 paper in the American Journal of Mathematics.

Statement of the Theorem

Let X *be an* n*-connected pointed space (a pointed CW-complex or pointed simplicial set). The map*

$$X \to \Omega(X \Omega S^1)$$

induces a map

$$\pi_k(X) \to \pi_k(X \Omega S^1))$$

on homotopy groups, where Ω denotes the loop functor and Ω denotes the smash product. The suspension theorem then states that the induced map on homotopy groups is an isomorphism if $k \leq 2n$ and an epimorphism if $k = 2n + 1$.

A basic result on loop spaces gives the relation

$$\pi_k(\Omega(X \Omega S^1)) \cong \pi_{k+1}(X \Omega S^1)$$

so the theorem could otherwise be stated in terms of the map

$$\pi_k(X) \to \pi_{k+1}(X \Omega S^1),$$

with the small caveat that in this case one must be careful with the indexing.

Corollary 1

Let S^n denote the n-sphere and note that it is $(n - 1)$-connected so that the groups $\pi_{n+k}(S^n)$ stabilise for

$$n \geq k + 2$$

by the Freudenthal theorem. These groups represent the kth stable homotopy group of spheres.

Corollary 2

More generally, for fixed $k \leq 1$, $k \leq 2n$ for sufficiently large n, so that any n-connected space X will have corresponding stabilised homotopy groups. These groups are actually the homotopy groups of an object corresponding to X in the stable homotopy category.

Hurewicz Theorem

In mathematics, the Hurewicz theorem is a basic result of algebraic topology, connecting homotopy theory with homology theory via a map known as the Hurewicz homomorphism. The theorem is named after Witold Hurewicz, and generalises earlier results from Henri Poincaré.

Statement of the Theorems

The Hurewicz theorems are a key link between homotopy groups and homology groups.

Absolute Version

For any space X and positive integer k there exists a group homomorphism

$$h_* : \pi_k(X) \to H_k(X)$$

called the Hurewicz homomorphism from the k-th homotopy group to the k-th homology group (with integer coefficients), which for $k = 1$ is equivalent to the canonical abelianisation map

$$h_* : \pi_1(X) \to \pi_1(X) / [\pi_1(X), \pi_1(X)].$$

The Hurewicz theorem states that if X is $(n - 1)$-connected, the Hurewicz map is an isomorphism for all $k \leq n$ when $n \geq 2$ and abelianisation for $n = 1$. In particular, this theorem says that the abelianisation of the first homotopy group (the fundamental group) is isomorphic to the first homology group:

$$H_1(X) \cong \pi_1(X) / [\pi_1(X), \pi_1(X)].$$

The first homology group therefore vanishes if X is path-connected and $\varphi_1(X)$ is a perfect group.

In addition, the Hurewicz homomorphism is an epimorphism from $\pi_{n+1}(X) \to H_{n+1}(X)$ whenever X is $(n - 1)$-connected, for $n \geq 2$.

The group homomorphism is given in the following way. Choose canonical generators $u_n \in H_n(S^n)$. Then a homotopy class of maps $f \in \pi_n(X)$ is taken to $f_*(u_n) \in H_n(X)$.

Relative Version

For any pair of spaces (X,A) and integer $k > 1$ there exists a homomorphism

$$h_* : \pi_k(X,A) \rightarrow H_k(X,A)$$

from relative homotopy groups to relative homology groups. The Relative Hurewicz Theorem states that if each of X, A are connected and the pair (X,A) is $(n\text{-}1)$-connected then $H_k(X,A) = 0$ for $k < n$ and $H_n(X,A)$ is obtained from $\varphi_n(X,A)$ by factoring out the action of $\varphi_1(A)$. This is proved in, for example, Whitehead (1978) by induction, proving in turn the absolute version and the Homotopy Addition Lemma.

This relative Hurewicz theorem is reformulated by Brown & Higgins (1981) as a statement about the morphism

$$\pi_n(X,A) \rightarrow \pi_n(X \cup CA).$$

This statement is a special case of a homotopical excision theorem, involving induced modules for n>2 (crossed modules if n=2), which itself is deduced from a higher homotopy van Kampen theorem for relative homotopy groups, whose proof requires development of techniques of a cubical higher homotopy groupoid of a filtered space.

Triadic Version

For any triad of spaces $(X;A,B)$ (i.e. space X and subspaces A,B) and integer $k > 2$ there exists a homomorphism

$$h_* : \pi_k(X;A,B) \rightarrow H_k(X;A,B)$$

from triad homotopy groups to triad homology groups. Note that $H_k(X;A,B) \cong H_k(X \cup (C(A \cup B))$.

The Triadic Hurewicz Theorem states that if X, A, B, and $C = A \cap B$ are connected, the pairs (A,C), (B,C) are respectively $(p\text{-}1)$-, $(q\text{-}1)$-connected, and the triad $(X;A,B)$ is $p+q\text{-}2$ connected, then $H_k(X;A,B) = 0$ for $k < p+q\text{-}2$ and $H_{p+q-1}(X;A)$ is obtained from $\pi_{p+q-1}(X;A,B)$ by factoring out the action of $\pi_1(A)\text{-}B)$ and the generalised Whitehead products.

The proof of this theorem uses a higher homotopy van Kampen type theorem for triadic homotopy groups, which requires a notion of the fundamental cat^n-group of an n-cube of spaces.

Simplicial Set Version

The Hurewicz theorem for topological spaces can also be stated for n-connected simplicial sets satisfying the Kan condition.

Rational Hurewicz Theorem

Rational Hurewicz theorem: Let X be a simply connected topological space with $\pi_i(X)\otimes\mathbb{Q}=0$ for $i\leq r$. Then the Hurewicz map

$$h\otimes\mathbb{Q}:\pi_i(X)\otimes\mathbb{Q}\to H_i(X;\mathbb{Q})$$

induces an isomorphism for $1\leq i\leq 2r$ and a surjection for $i = 2r + 1$.

Künneth Theorem

In mathematics, especially in homological algebra and algebraic topology, a Künneth theorem is a statement relating the homology of two objects to the homology of their product. The classical statement of the Künneth theorem relates the singular homology of two topological spaces X and Y and their product space $X\times Y$. In the simplest possible case the relationship is that of a tensor product, but for applications it is very often necessary to apply certain tools of homological algebra to express the answer.

A Künneth theorem or Künneth formula is true in many different homology and cohomology theories, and the name has become generic. These many results are named for the German mathematician Hermann Künneth.

Singular Homology with Coefficients in a Field

Let X and Y be two topological spaces. In general one uses singular homology; but if X and Y happen to be CW complexes, then this can be replaced by cellular homology, because that is isomorphic to singular homology. The simplest case is when the coefficient ring for homology is a field F. In this situation, the Künneth theorem (for singular homology) states that for any integer k,

$$\bigoplus_{i+j=k} H_i(X;F)\otimes H_j(Y;F)\cong H_k(X\times Y;F).$$

Furthermore, the isomorphism is a natural isomorphism. The map from the sum to the homology group of the product is called the *cross product*. More precisely, there is a cross product operation by which an i-cycle on X and a j-cycle on Y can be combined to create an $(i+j)$-cycle on $X\times Y$; so that there is an explicit linear mapping defined from the direct sum to $H_k(X\times Y)$.

A consequence of this result is that the Betti numbers, the dimensions of the homology with Q coefficients, of $X\times Y$ can be determined from those of X and Y. If $p_Z(t)$ is the generating function of the sequence of Betti numbers $b_k(Z)$ of a space Z, then

$$p_{X\times Y}(t)=p_X(t)p_Y(t).$$

Here when there are finitely many Betti numbers of X and Y, each of which is a natural number rather than “, this reads as an identity on Poincaré polynomials. In the general case these are formal power series with possibly infinite coefficients, and have to be interpreted accordingly. Furthermore, the above statement holds not only for the Betti numbers but also for the generating functions of the dimensions of the homology over any field. (If the integer homology is not torsion-free then these numbers may differ from the standard Betti numbers.)

Singular Homology with Coefficients in a PID

The above formula is simple because vector spaces over a field have very restricted behaviour. As the coefficient ring becomes more general, the relationship becomes more complicated. The next simplest case is the case when the coefficient ring is a principal ideal domain. This case is particularly important because the integers $\mathbb{Z}$ are a PID.

In this case the equation above is no longer always true. A correction factor appears to account for the possibility of torsion phenomena. This correction factor is expressed in terms of the Tor functor, the first derived functor of the tensor product.

When R *is a PID, then the correct statement of the Künneth theorem is that for any topological spaces* X *and* Y *there are natural short exact sequences*

$$0 \to \bigoplus_{i+j=k} H_i(X;R) \otimes_R H_j(Y;R) \to H_k(X \times Y;R) \to \bigoplus_{i+j=k-1} \mathrm{Tor}_1^R(H_i(X;R), H_j(Y;R)) \to 0.$$

Furthermore these sequences split, but not canonically.

Example

The short exact sequences just described can easily be used to compute the homology groups

$$H_k(\mathbb{RP}^2 \times \mathbb{RP}^2; \mathbb{Z})$$

with integer coefficients of the product $RP^2 \ x \ RP^2$ of two real projective planes. These spaces are CW complexes. Denoting the homology group $H_i(\mathbb{RP}^2;\mathbb{Z})$ by h_i for brevity's sake, one knows from a simple calculation with cellular homology that

$$h_0 \cong \mathbb{Z}, \qquad h_1 \cong \mathbb{Z}/(2)$$

and h_i is zero for all other values of i. The only non-zero Tor group (torsion product) which can be formed from these values of h_i is

$$\mathrm{Tor}_1^{\mathbb{Z}}(h_1, h_1) \cong \mathrm{Tor}_1^{\mathbb{Z}}(\mathbb{Z}/(2), \mathbb{Z}/(2)) \cong \mathbb{Z}/(2).$$

Therefore the Künneth short exact sequence reduces in every degree to an isomorphism, because there is a zero group in each case on either the left or the right side in the sequence. The result is

$$
\begin{aligned}
H_0(\mathbb{RP}^2 \times \mathbb{RP}^2;\mathbb{Z}) &\cong h_0 \otimes h_0 \cong \mathbb{Z} \\
H_1(\mathbb{RP}^2 \times \mathbb{RP}^2;\mathbb{Z}) &\cong h_0 \otimes h_1 \oplus h_1 \otimes h_0 \cong \mathbb{Z}/(2) \oplus \mathbb{Z}/(2) \\
H_2(\mathbb{RP}^2 \times \mathbb{RP}^2;\mathbb{Z}) &\cong h_1 \otimes h_1 \cong \mathbb{Z}/(2) \\
H_3(\mathbb{RP}^2 \times \mathbb{RP}^2;\mathbb{Z}) &\cong \mathrm{Tor}_1^{\mathbb{Z}}(h_1, h_1) \cong \mathbb{Z}/(2)
\end{aligned}
$$

and all the other homology groups are zero.

The Künneth Spectral Sequence

For a general commutative ring R, *the homology of* X *and* Y *is related to the homology of their product by a Künneth spectral sequence*

$$E^2_{pq} = \bigoplus_{q_1+q_2=q} \mathrm{Tor}_p^R(H_{q_1}(X;R), H_{q_2}(Y;R)) \Rightarrow H_{p+q}(X \times Y;R).$$

In the cases described above, this spectral sequence collapses to give an isomorphism or a short exact sequence.

Relation with Homological Algebra, and Idea of Proof

The chain complex of the space $X \times Y$ *is related to the chain complexes of* X *and* Y *by a natural quasi-isomorphism*

$$C_*(X \times Y) \cong C_*(X) \otimes C_*(Y).$$

For singular chains this is the theorem of Eilenberg and Zilber. For cellular chains on CW complexes, it is a straightforward isomorphism. Then the homology of the tensor product on the right is given by the spectral Künneth formula of homological algebra.

The freeness of the chain modules means that in this geometric case it is not necessary to use any hyperhomology or total derived tensor product.

There are analogues of the above statements for singular cohomology and sheaf cohomology. For sheaf cohomology on an algebraic variety, Grothendieck found six spectral sequences relating the possible hyperhomology groups of two chain complexes of sheaves and the hyperhomology groups of their tensor product.

Künneth Theorems in Generalised Homology and Cohomology Theories

There are many generalised or extraordinary homology and cohomology theories for topological spaces. K-theory and cobordism are the best-known. Their striking common feature (not their definition) is that they do not arise from ordinary chain complexes. Thus Künneth theorems can not be obtained by the above methods of homological algebra.

Nevertheless Künneth theorems in just the same form have been proved in very many cases by various other methods. The first were Atiyah's Künneth theorem for complex K-theory and Conner and Floyd's result in cobordism.

A general method of proof emerged, based upon a homotopical theory of modules over highly structured ringspectra. The homotopy category of such modules closely resembles the derived categories of homological algebra.

Poincaré Duality

In mathematics, the Poincaré duality theorem named after Henri Poincaré, is a basic result on the structure of the homology and cohomology groups of manifolds. It states that if M is an n-dimensional oriented closed manifold (compact and without boundary), then the kth cohomology group of M is isomorphic to the $(n - k)$th homology group of M, for all integers k

$$H^k(M) \cong H_{n-k}(M).$$

Poincaré duality holds for any coefficient ring, so long as one has taken an orientation with respect to that coefficient ring; in particular, since every manifold has a unique orientation mod 2, Poincaré duality holds mod 2 without any assumption of orientation.

History

A form of Poincaré duality was first stated, without proof, by Henri Poincaré in 1893. It was stated in terms of Betti numbers: The kth and (n - k) th Betti numbers of a closed (i.e. compact and without boundary) orientable n-manifold are equal. The cohomology concept was at that time about 40 years from being clarified. In his 1895 paper Analysis Situs, Poincaré tried to prove the theorem using topological intersection theory, which he had invented. Criticism of his work by Poul Heegaard led him to realise that his proof was seriously flawed. In the first two complements to Analysis Situs, Poincaré gave a new proof in terms of dual triangulations.

Poincaré duality did not take on its modern form until the advent of cohomology in the 1930s, when Eduard Èech and Hassler Whitney invented the cup and cap products and formulated Poincaré duality in these new terms.

Modern Formulation

The modern statement of the Poincaré duality theorem is in terms of homology and cohomology: if Mis a closed oriented n-manifold, and k is an integer, then there is a canonically defined isomorphism from the k-th homology group $H_k(M)$ to the (n - k)th cohomology group $H^{n-k}(M)$. (Here, homology and cohomology is taken with coefficients in the ring of integers, but the isomorphism holds for any coefficient ring.) Specifically, one maps an element of $H^k(M)$ to its cap product with a fundamental class of M, which will exist for oriented M.

For non-compact oriented manifolds, one has to replace cohomology by cohomology with compact support.

Homology and cohomology groups are defined to be zero for negative degrees, so Poincaré duality in particular implies that the homology and cohomology groups of orientable closed n-manifolds are zero for degrees bigger than n.

Dual Cell Structures

Given a triangulated manifold, there is a corresponding dual polyhedral decomposition. The dual polyhedral decomposition is a cell decomposition of the manifold such that the k-cells of the dual polyhedral decomposition are in bijective correspondence with the n-k-cells of the triangulation, generalising the notion of dual polyhedra.

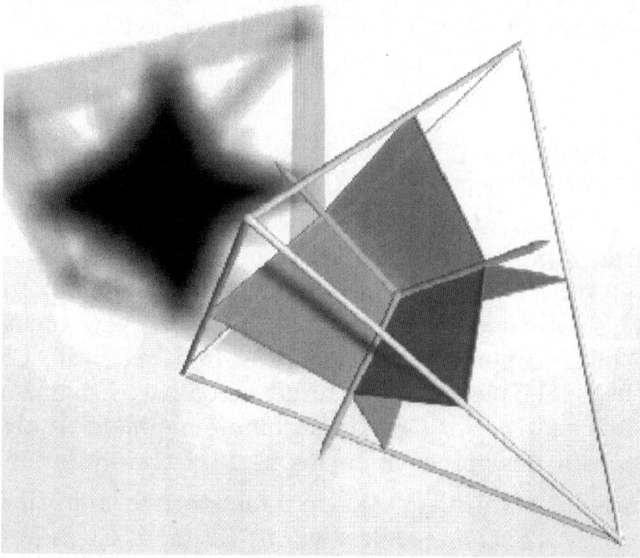

Figure 4: $\cup_{S\in T}\Delta \cap DS$ — a picture of the parts of the dual-cells in a top-dimensional simplex.

Precisely, let T be a triangulation of an n-manifold M. Let S be a simplex of T. We denote the dual cell (to be defined precisely) corresponding to S by DS. Let Δ be a top-dimensional simplex of T containing S. So we can think of S as a subset of the vertices of Δ.

Then $\Delta \cap DS$ is defined to be the convex hull (in Δ) of the barycentres of all subsets of the vertices of Δ that contain S. One can check that if S is i-dimensional, then DS is an n-i-dimensional cell. Moreover, the dual cells to T form a CW-decomposition of M, and the only n-i-dimensional dual cell that intersects an i-cell S is DS. Thus the pairing $C_i M \otimes C_{n-i} M \to \mathbb{Z}$ given by taking intersections induces an isomorphism $C_i M \to C^{n-i} M$, where here C_i is the cellular homology of the triangulation T, and $C_{n-i} M$ and $C^{n-i} M$ are the cellular homologies and cohomologies of the dual polyhedral/CW decomposition the manifold respectively. The fact that this is an isomorphism of chain complexes is a proof of Poincaré Duality. Roughly speaking, this amounts to the fact that the boundary relation for the triangulation T is the incidence relation for the dual polyhedral decomposition under the correspondence $S \mapsto DS$.

Naturality

Note that H^k is a contravariant functor while H_{n-k} is covariant. The family of isomorphisms

$$D_M : H^k(M) \to H_{n-k}(M)$$

is natural in the following sense: if

$$f : M \to N$$

is a continuous map between two oriented n-manifolds which is compatible with orientation, i.e. which maps the fundamental class of M to the fundamental class of N, then

$$D_N = f_* D_M f^*,$$

where f_* and f^* are the maps induced by f in homology and cohomology, respectively.

Note the very strong and crucial hypothesis that f maps the fundamental class of M to the fundamental class of N. Naturality does not hold for an arbitrary continuous map f, since in general f^* is not an injection on cohomology. For example if f is a covering map then it maps the fundamental class of M to a multiple of the fundamental class of N. This multiple is the degree of the map f.

Bilinear Pairings Formulation

Assuming M *is compact boundaryless and orientable, let* $\tau H_i M$ *denote the torsion subgroup of* $H_i M$ *and let* $fH_i M = H_i M / \tau H_i M$ *be the free part – all homology groups taken with integer coefficients in this section. Then there are bilinear maps which are duality pairings (explained below).*

$$fH_i M \otimes fH_{n-i} M \to \mathbb{Z}$$

and

$$\tau H_i M \otimes \tau H_{n-i-1} M \to \mathbb{Q}/\mathbb{Z}.$$

(Here $\mathbb{Q}/\mathbb{Z}$ is the quotient of the rationals by the integers, taken as an additive group.)

(Notice that in the torsion linking form, there is a -1 in the dimension, so the paired dimensions add up to n - 1, rather than to n.)

The first form is typically called the *intersection product* and the 2nd the *torsion linking form. Assuming* the manifold M is smooth, the intersection product is computed by perturbing the homology classes to be transverse and computing their oriented intersection number. For the torsion linking form, one computes the pairing of x and y by realising nx as the boundary of some class z. The form is the fraction with numerator the transverse intersection number of z with y and denominator n.

The statement that the pairings are duality pairings means that the adjoint maps

$$fH_i M \to \mathrm{Hom}_{\mathbb{Z}}(fH_{n-i} M, \mathbb{Z})$$

and

$$\tau H_i M \to \mathrm{Hom}_{\mathbb{Z}}(\tau H_{n-i-1} M, \mathbb{Q}/\mathbb{Z})$$

are isomorphisms of groups.

This result is an application of Poincaré Duality $H_i M \simeq H^{n-i} M$ together with the Universal coefficient theorem which gives an identification $fH^{n-i} M \equiv \mathrm{Hom}(H_{n-i} M; \mathbb{Z})$ and $\tau H^{n-i} M \equiv \mathrm{Ext}(H_{n-i-1} M; \mathbb{Z}) \equiv \mathrm{Hom}(\tau H_{n-i-1} M; \mathbb{Q}/\mathbb{Z})$. Thus, Poincaré duality says that $fH_i M$ and $fH_{n-i} M$ are isomorphic, although there is no natural map giving the isomorphism, and similarly $\tau H_i M$ and $\tau H_{n-i-1} M$ are also isomorphic, though not naturally.

This approach to Poincaré duality was used by Przytycki and Yasuhara to give an elementary homotopy and diffeomorphism classification of 3-dimensional lens spaces.

Thom Isomorphism Formulation

Poincaré Duality is closely related to the Thom Isomorphism Theorem, as we will explain here. For this exposition, let *M* be a compact, boundaryless oriented n-manifold. Let *M*×*M* be the product of *M* with itself, let *V* be an open tubular neighbourhood of the diagonal in *M*×*M*. Consider the maps:

- $H_*M \otimes H_*M \to H_*(M \times M)$ the Homology Cross Product
- $H_*(M \times M) \to H_*\left(M \times M, (M \times M) \setminus V\right)$ inclusion.
- $H_*\left(M \times M, (M \times M) \setminus V\right) \to H_*(\nu M, \partial \nu M)$ excision map where í*M* is the normal disc bundle of the diagonal in *M*×*M*.
- $H_*(\nu M, \partial \nu M) \to H_{*-n}M$ the Thom Isomorphism. This map is well-defined as there is a standard identification $\nu M \equiv TM$ which is an oriented bundle, so the Thom Isomorphism applies.

Combined, this gives a map $H_iM \otimes H_jM \to H_{i+j-n}M$, which is the *intersection product* — strictly speaking it is a generalisation of the intersection product above, but it is also called the intersection product. A similar argument with the Künneth theorem gives the *torsion linking form*.

This formulation of Poincaré Duality has become quite popular as it provides a means to define Poincaré Duality for any generalised homology theories provided one has a Thom Isomorphism for that homology theory. A Thom isomorphism theorem for a homology theory is now accepted as the generalised notion of orientability for a homology theory. For example, a *spin*^{c}-structure on a manifold turns out to be precisely what is needed to be orientable in the sense of complex topological k-theory.

Generalisations and Related Results

The Poincaré-Lefschetz duality theorem is a generalisation for manifolds with boundary. In the non-orientable case, taking into account the sheaf of local orientations, one can give a statement that is independent of orientability.

Blanchfield duality is a version of Poincaré duality which provides an isomorphism between the homology of an abelian covering space of a manifold and the corresponding cohomology with compact supports. It is used to get basic structural results about the Alexander module and can be used to define the signatures of a knot.

With the development of homology theory to include K-theory and other extraordinary theories from about 1955, it was realised that the homology H_* could be replaced by other theories, once the products on manifolds were constructed; and there are now textbook treatments in generality. More specifically, there is a general Poincaré duality theorem for generalised homology theories which requires a notion of orientation with respect to a homology theory, and is formulated in terms of a generalised Thom Isomorphism Theorem. The Thom Isomorphism Theorem in this regard can be considered as the germinal idea for Poincaré duality for generalised homology theories.

Verdier duality is the appropriate generalisation to (possibly singular) geometric objects, such as analytic spaces or schemes, while intersection homology was developed R. MacPherson and M. Goresky for stratified spaces, such as real or complex algebraic varieties, precisely so as to generalise Poincaré duality to such stratified spaces.

There are many other forms of geometric duality in algebraic topology, including Lefschetz duality, Alexander duality, Hodge duality, and S-duality.

Universal Coefficient Theorem

In mathematics, the universal coefficient theorem in algebraic topology establishes the relationship in homology theory between the *integral homology* of a topological space *X*, and its *homology with coefficients* in any abelian group *A*. It states that the integral homology groups

$$H_i(X,\mathbb{Z})$$

completely determine the groups

$$H_i(X,A)$$

Here H_i might be the simplicial homology or more general singular homology theory: the result itself is a pure piece of homological algebra about chain complexes of free abelian groups. The form of the result is that other coefficients *A* may be used, at the cost of using a Tor functor.

For example it is common to take A to be $\mathbb{Z}/2\mathbb{Z}$, so that coefficients are modulo 2. This becomes straightforward in the absence of 2-torsion in the homology. Quite generally, the result indicates the relationship that holds between the Betti numbers b_i of X and the Betti numbers $b_{i,F}$ with coefficients in a field F. These can differ, but only when the characteristic of F is a prime number p for which there is some p-torsion in the homology.

Statement

Consider the tensor product $H_i(X,\mathbb{Z}) \otimes A$. *The theorem states that there is an injective group homomorphism é from this group to* $H_i(X,A)$, *which has cokernel* $\mathrm{Tor}(H_{i-1}(X,\mathbb{Z}),A)$.

In other words, there is a natural short exact sequence

$$0 \to H_i(X,\mathbb{Z}) \otimes A \to H_i(X,A) \to \mathrm{Tor}(H_{i-1}(X,\mathbb{Z}),A) \to 0.$$

Furthermore, this is a split sequence (but the splitting is not *natural).*

The Tor group on the right can be thought of as the obstruction to é being an isomorphism.

Universal Coefficient Theorem for Cohomology

There is also a universal coefficient theorem for cohomology involving the Ext functor, stating that there is a natural short exact sequence

$$0 \to \mathrm{Ext}(H_{i-1}(X,\mathbb{Z}),A) \to H^i(X,A) \to \mathrm{Hom}(H_i(X,\mathbb{Z}),A) \to 0.$$

As in the homological case, the sequence splits, though not naturally.

Example: Mod 2 Cohomology of the Real Projective Space

Let $X = RP^n$, *the real projective space. We compute the singular cohomology of* X *with coefficients in* $R := Z_2$.

knowing that the integer homology is given by:

$$H_i(X;Z) = \begin{cases} Z & i = 0 \text{ or } i = n \text{ odd}, \\ Z/2Z & 0 < i < n, i \text{ odd}, \\ 0 & \text{else}. \end{cases}$$

We have $\mathrm{Ext}(R,R) = R, \mathrm{Ext}(Z,R) = 0$, so that the above exact sequences yield

$$\forall i = 0\ldots n, H^i(X;R) = R.$$

In fact the total cohomology ring structure is

$$H^*(X;R) = R[w] / < w^{n+1} > .$$

Seifert–van Kampen Theorem

In mathematics, the Seifert-van Kampen theorem of algebraic topology, sometimes just called van Kampen's theorem, expresses the structure of the fundamental group of a topological space X, in terms of the fundamental groups of two open, path-connected subspaces U and V that cover X. It can therefore be used for computations of the fundamental group of spaces that are constructed out of simpler ones.

The underlying idea is that paths in X can be partitioned into journeys: through the intersection W of U and V, through U but outside V, and through V outside U. In order to move segments of paths around, by homotopy to form loops returning to a base point w in W, we should assume U, V and W are path-connected and that W isn't empty. We also assume that U and V are open subspaces with union X.

Equivalent Formulations

In the language of combinatorial group theory, $\varphi_1(X,w)$ is the free product with amalgamation of $\varphi_1(U,w)$ and $\varphi_1(V,w)$, with respect to the (not necessarily injective) homomorphisms I *and* J. *Given group presentations:*

$$\pi_1(U,w) = \langle u_1,\dots,u_k \mid \alpha_1,\dots,\alpha_l \rangle$$

$$\pi_1(V,w) = \langle v_1,\dots,v_m \mid \beta_1,\dots,\beta_n \rangle, \text{ and}$$

$$\pi_1(W,w) = \langle w_1,\dots,w_p \mid \gamma_1,\dots,\gamma_q \rangle$$

the amalgamation can be presented as

$$\pi_1(X,w) = \langle u_1,\dots,u_k, v_1,\dots,v_m \mid \alpha_1,\dots,\alpha_l, \beta_1,\dots,\beta_n, I(w_1)J(w_1)^{-1},\dots,I(w_p)J(w_p)^{-1} \rangle.$$

In category theory, $\varphi_1(X,w)$ is the pushout, in the category of groups, of the diagram:

$$\pi_1(U,w) \leftarrow \pi_1(W,w) \rightarrow \pi_1(V,w).$$

Van Kampen's Theorem for Fundamental Groups

Van Kampen's theorem for fundamental groups:

Let X be a topological space which is the union of two open and path connected subspaces U_1, U_2. Suppose $U_1 \cap U_2$ is path connected and let x_0 be a point in it that will be used as the base of all fundamental *groups. Then* X *is path connected and the inclusion morphisms draw a commutative pushout diagram:*

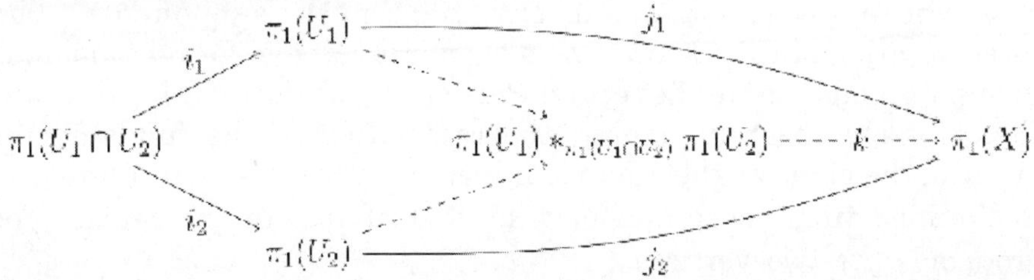

the natural morphism k is an isomorphism, that is, the fundamental group of X is the free product of the fundamental groups of U_1 and U_2 with amalgamation *of* $\pi_1(U_1 \cap U_2, x_0)$.

Usually the morphisms induced by inclusion in this theorem are not themselves injective, and the more precise version of the statement is in terms of pushouts of groups. The notion of pushout in the category of groupoids allows for a version of the theorem for the non path connected case, using the fundamental groupoid $\varphi_1(X,A)$ on a set A of base points,. *This groupoid consists of homotopy classes relative to the end points of paths in X joining points of* $A \cap X$. *In particular, if X is a contractible space, and A consists of two distinct points of X, then* $\varphi_1(X,A)$ *is easily seen to be isomorphic to the groupoid often written* $\mathcal{I}$ *with two vertices and exactly one morphism between any two vertices. This groupoid plays a role in the theory of groupoids analogous to that of the group of integers in the theory of groups*. The groupoid $\mathcal{I}$ also allows for groupoids a notion of homotopy: it is a *unitinterval object* in the category of groupoids.

Theorem: *Let the topological space X be covered by the interiors of two subspaces* X_1, X_2 *and let A be a set which meets each path component of* X_1, X_2 *and* $X_0 := X_1 \cap X_2$. *Then A meets each path component of X and the diagram P of morphisms induced by inclusion*

$$\begin{array}{ccc} \pi_1(X_0, A) & \xrightarrow{\pi_1(i_1)} & \pi_1(X_1, A) \\ {\scriptstyle \pi_1(i_2)}\downarrow & & \downarrow{\scriptstyle \pi_1(j_1)} \\ \pi_1(X_2, A) & \xrightarrow[\pi_1(j_2)]{} & \pi_1(X, A) \end{array}$$

is a pushout diagram in the category of groupoids.

One can easily find cases where X is connected but is the union of the interiors of two subspaces, each with say 402 path components and whose intersection has say 1004 path components. The interpretation of this theorem as a calculational tool for fundamental groups needs some development of 'combinatorial groupoid theory',. This theorem implies the calculation of the fundamental group of the circle as the group of integers, since the group of integers is obtained from the groupoid $\mathcal{I}$ by identifying, in the category of groupoids, its two vertices.

There is a version of the last theorem when X is covered by the union of the interiors of a family $\{U_\lambda : \lambda \in \Lambda\}$ of subsets. The conclusion is that if A meets each path component of all 1,2,3-fold intersections of the sets U_λ, then A meets all path components of X and the diagram $\bigsqcup_{(\lambda,\mu)\in\Lambda^2} \pi_1(U_\lambda \cap U_\mu, A) \rightrightarrows \bigsqcup_{\lambda\in\Lambda} \pi_1(U_\lambda, A) \to \pi_1(X, A)$ of morphisms induced by inclusions is a coequaliser in the category of groupoids.

Examples

One can use Van Kampen's theorem to calculate fundamental groups for topological spaces that can be decomposed into simpler spaces. For example, consider the sphere S^2. *Pick open sets* $A = S^2$ "n*and* $B = S^2$ " s *where n and s denote the north and south poles respectively. Then we have the property that A, B and* $A \cap B$ *are open path connected sets. There is a commutative diagram including* $A \cap B$ *into A and B and then another inclusion from A and B into* S^2 *and that there is a corresponding diagram of homomorphisms between the fundamental groups of each subspace. Applying Van Kampen's theorem gives the result* $\varphi_1(S^2) = \varphi_1(A) * \varphi_1(B) / \ker(\Phi)$. *However A and B are both homeomorphic to* $\mathbb{R}^2$ *which is simply connected, so both A and B have trivial fundamental groups. It is clear from this that the fundamental group of* S^2 *is trivial.*

A more complicated example is the calculation of the fundamental group of a genus *n* orientable surface *S*, otherwise known as the *genus n surface group*. One can construct *S* using its standard fundamental polygon. For the first open set *A*, pick a disk within the centre of the polygon. Pick *B* to be the complement in *S* of the centre point of *A*. Then the intersection of *A* and *B* is an annulus, which is known to be homotopy equivalent to (and so has the same fundamental group as) a circle. Then $\pi_1(A \cap B) = \pi_1(S^1)$, which is the integers, and $\varphi_1(A)$

$= \varphi_1(D^2) = 1$. Thus the inclusion of $\pi_1(A \cap B)$ into $\varphi_1(A)$ sends any generator to the trivial element. However, the inclusion of $\pi_1(A \cap B)$ into $\varphi_1(B)$ is not trivial. In order to understand this, first one must calculate $\varphi_1(B)$. This is easily done as one can deformation retract B (which is S with one point deleted) onto the edges labelled by $A_1B_1A_1^{-1}B_1^{-1}A_2B_2A_2^{-1}B_2^{-1}\ldots A_nB_nA_n^{-1}B_n^{-1}$. This space is known to be the wedge sum of 2n circles (also called a bouquet of circles), which further is known to have fundamental group isomorphic to the free group with $2n$ generators, which in this case can be represented by the edges themselves: $\{A_1, B_1, \ldots, A_n, B_n\}$. We now have enough information to apply Van Kampen's theorem. The generators are the loops $\{A_1, B_1, \ldots, A_n, B_n\}$ (A is simply connected, so it contributes no generators) and there is exactly one relation: $A_1B_1A_1^{-1}B_1^{-1}A_2B_2A_2^{-1}B_2^{-1}\ldots A_nB_nA_n^{-1}B_n^{-1} = 1$. Using generators and relations, this group is denoted

$$\langle A_1, B_1, \ldots, A_n, B_n \mid A_1B_1A_1^{-1}B_1^{-1}\ldots A_nB_nA_n^{-1}B_n^{-1}\rangle.$$

Generalisations

This theorem has been extended to the non-connected case by using the fundamental groupoid $_1$(X,A) on a set A of base points, which consists of homotopy classes of paths in X joining points of X which lie in A. The connectivity conditions for the theorem then become that A meets each path-component of U,V,W. The pushout is now in the category of groupoids. This extended theorem allows the determination of the fundamental group of the circle, and many other useful cases. For example, if the intersection W has two path components, it is convenient to let A consist of one point in each of these components. A theorem for arbitrary covers, with the restriction that A meets all threefold intersections of the sets of the cover, is given in the paper by Brown and Razak cited below. Applications of the fundamental groupoid on a set of base points to the Jordan curve theorem, Covering space, and orbit space are given in Ronald Brown's book cited below.

In the case of orbit spaces, it is convenient to take A to include all the fixed points of the action. An example here is the conjugation action on the circle.

The version that allows more than two overlapping sets but with A a singleton is also given in Allen Hatcher's book below, theorem 1.20.

In fact, we can extend van Kampen's theorem significantly further by considering the fundamental groupoid Đ(X), a small category whose objects are points of X and whose arrows are homotopy equivalences of paths. In this case, to determine the fundamental groupoid of a space, we need only know the fundamental groupoids of a covering of the space by path-connected components: create a new category in which the objects are fundamental groupoids of the path-connected open sets that form the cover, with an arrow between groupoids if the domain space is a subspace of the codomain. Then van Kampen's theorem is the assertion that the fundamental groupoid of X is the colimit of the diagram. One can think of the colimit relation as a disjoint union followed by a quotient. A precise statement of the theorem along with proof is given in Peter May's book A Concise Introduction to Algebraic Topology, chapter 2.

References to higher dimensional versions of the theorem which yield some information on homotopy types are given in an article on higher dimensional group theories and groupoids.

Fundamental groups also appear in algebraic geometry and are the main topic of Alexander Grothendieck's first Séminaire de géométrie algébrique (SGA1). A version of van Kampen's theorem appears there, and is proved along quite different lines than in algebraic topology, namely descent theory. A similar proof works in algebraic topology .

Whitehead Theorem

In homotopy theory (a branch of mathematics), the Whitehead theorem states that if a continuous mapping *f* between topological spaces *X* and *Y* induces isomorphisms on all homotopy groups, then *f* is a homotopy equivalence provided *X* and *Y* are connected and have the homotopy-type of CW complexes. This result was proved by J. H. C. Whitehead in two landmark papers from 1949, and provides a justification for working with the CW complex concept that he introduced there.

Statement

More accurately, we suppose given CW complexes X *and* Y, *with respective base points* x *and* y. *Given a continuous mapping*

$$f : X \to Y$$

such that $f(x) = y$, we consider for $n \geq 0$ the induced homomorphisms

$$f_* : \pi_n(X,x) \to \pi_n(Y,y),$$

where φ_n denotes for $n \geq 1$ the n-th homotopy group. For $n = 0$ this means the mapping of the path-connected components; if we assume both *X* and *Y* are connected we can ignore this as containing no information. We say that *f* is a weak homotopy equivalence if the homomorphisms f_* are all isomorphisms. The Whitehead theorem then states that a weak homotopy equivalence, for connected CW complexes, is a homotopy equivalence.

Spaces with Isomorphic Homotopy Groups may not be Homotopy Equivalent

A word of caution: it is not enough to assume $\varphi_n(X)$ *is isomorphic to* $\varphi_n(Y)$ *for each* $n \geq 1$ *in order to conclude that* X *and* Y *are homotopy equivalent. One really needs a map* $f : X \rightarrow Y$ *inducing such isomorphisms in homotopy. For instance, take* $X= S^2 \times RP^3$ *and* $Y= RP^2 \times S^3$*. Then* X *and* Y *have the same fundamental group, namely* Z_2*, and the same universal cover, namely* $S^2 \times S^3$*; thus, they have isomorphic homotopy groups. On the other hand their homology groups are different (as can be seen from the Künneth formula); thus,* X *and* Y *are not homotopy equivalent.*

The Whitehead theorem does not hold for general topological spaces or even for all subspaces of R^n. For example, the Warsaw circle, a subset of the plane, has all homotopy groups zero, but the map from the Warsaw circle to a single point is not a homotopy equivalence. The study of possible generalisations of Whitehead's theorem to more general spaces is part of the subject of shape theory.

Generalisation to Model Categories

In any model category, a weak equivalence between cofibrant-fibrant objects is a homotopy equivalent.

Chapter 6

Some Elements of Algebra

Variable

A variable is a value that may change within the scope of a given problem or set of operations. In contrast, a constant is a value that remains unchanged, though often unknown or undetermined. The concepts of constants and variables are fundamental to many areas of mathematics and its applications. A "constant" in this context should not be confused with a mathematical constant which is a specific number independent of the scope of the given problem.

Dependent and Independent Variables

Variables are further distinguished as being either a dependent variable or an independent variable. Independent variables are regarded as inputs to a system and may take on different values freely. Dependent variables are those values that change as a consequence of changes in other values in the system.

When one value is completely determined by another, or several others, then it is called a function of the other value or values. In this case the value of the function is a dependent variable and the other values are independent variables. The notation $f(x)$ is used for the value of the function f with x representing the independent variable. Similarly, notation such as $f(x, y, z)$ may be used when there are several independent variables.

What it Means for a Variable to Vary

Varying, in the context of mathematical variables, does not mean change in the course of time, but rather dependence on the context in which the variable is used. This can be the immediate context of the expression in which the variable occurs, as in the case of summation variables or variables that designate the argument of a function being defined. The context can also be larger, for instance when a variable is used to designate a value occurring in a hypothesis of the discussion at hand. In some cases nothing varies at all, and alternative names can be used instead of "variable": a parameter is a value that is fixed in the statement of the problem being studied (although its value may not be explicitly known), an unknown is a variable that is introduced to stand for a constant value that is not initially known, but which may become known by solving some equation(s) for it, and an indeterminate is a symbol that need not stand for anything else but is an abstract value in itself.

In all these cases the term "variable" is often still used because the rules for the manipulation of these symbols are the same; however, in some languages other than English, one distinguishes between "variables" in functions and "unknown quantities" in equations ("incógnita" in Portuguese/Spanish, "inconnue" in French).

Examples

If one defines a function *f* from the real numbers to the real numbers by

$$f(x) = x^2 + \sin(x+4)$$

then x is a variable standing for the argument of the function being defined, which can be any real number. In the identity

$$\sum_{i=1}^{n} i = \frac{n^2+n}{2}$$

the variable i is a summation variable which designates in turn each of the integers 1, 2, ..., n (it is also called index because its variation is over a discrete set of values) while n is a parameter (it does not vary within the formula).

In the theory of polynomials, a polynomial of degree 2 is generally denoted as $ax^2 + bx + c$, where a, $b$and c are called coefficients (they are assumed to be fixed, i.e., parameters of the problem considered) while x is called a variable. When studying this polynomial for its polynomial function this x stands for the function argument. When studying the polynomial as an object in itself, x is taken to be an indeterminate, and would often be written with a capital letter instead to indicate this status.

Formulas from physics such as $E = mc^2$ or $PV = nRT$ (the ideal gas law) do not involve the mathematical notion of a variable, because the quantities E, m, P, V, n, and T are instead used to designate certain properties (energy, mass, pressure, volume, quantity, temperature) of the physical system.

Notation

In mathematics, single-symbol names for variables are the norm, with letter at the beginning of the alphabet, e.g. a, b, c commonly used for constants and letters at the end of the alphabet, e.g. x, y, z, and t commonly used for variables. In written mathematics, variables and constants are usually set in an italic type face.

Specific branches and applications of mathematics usually have specific naming conventions for variables. Variables with similar roles or meanings are often assigned consecutive letters. For example, the three axes in 3D coordinate space are conventionally called x, y, and z, while random variables in statistics are usually named X, Y, Z. In physics, the names of variables are largely determined by the physical quantity they describe, but various naming conventions exist.

A convention sometimes followed in statistics is to use X, Y, Z for the names of random variables, with these being replaced by x, y, z for observations or sample outcomes of those random variables. Another convention sometimes used in statistics is to denote population values of particular statistics by lower (or upper) case Greek letters, with sample-based estimates of those quantities being denoted by the corresponding lower (or upper) case letters from the ordinary alphabet.

General Introduction

Variables are used in open sentences. For instance, in the formula $x + 1 = 5$, x is a variable which represents an "unknown" number. Variables are often represented by Greek or Roman letters and may be used with other special symbols.

In mathematics, variables are essential because they let quantitative relationships to be stated in a general way. If we were forced to use actual values, then the relationships would only apply in a more narrow set of situations. For example:

State a mathematical definition for finding the number twice that of ANY other finite number:

$$2(x) = x + x \text{ or } x * 2$$

Now, all we need to do to find the double of a number is replace x with any number we want.

- $2(1) = 1 + 1 = 2$ or $1 * 2$
- $2(3) = 3 + 3 = 6$ or $3 * 2$
- $2(55) = 55 + 55 = 110$ or $55 * 2$
- etc.

So in this example, the variable x is a "placeholder" for any number—that is to say, a variable. One important thing we assume is that the value of x does not change, even though we do not know what xis. But in some algorithms, obviously, will change x, and there are various ways to then denote if we mean its old or new value—again, generally not knowing either, but perhaps (for example) that one is less than the other.

Naming Conventions

Mathematics has many conventions. Below are some of the more common. Many of the symbols have other conventional uses, but they may actually represent a constant or a specific function rather than a variable.

- a_i is often used to denote a term of a sequence.
- *a*, *b*, *c*, and *d* (sometimes extended to *e* and *f*) usually play similar roles or are made to represent parallel notions in a mathematical context. They often represent constants.

o The coefficients in an equation, for example the general expression of a polynomial or a Diophantine equation are often *a*, *b*, *c*, *d*, *e*, and *f*.

- *f* and *g* (sometimes *h*) commonly denote functions.
- *i*, *j*, and *k* (sometimes *l* or *h*) are often used as subscripts to denote indices.
- *l* and *w* are often used to represent the length and width of a figure.
- *m* and *n* usually denote integers and usually play similar roles or are made to represent parallel notions in a mathematical context, such a pair of dimensions.
    - o *n* commonly denotes a count of objects, or, in statistics, the number of individuals or observations.
- *p*, *q*, and *r* usually play similar roles or are made to represent parallel notions in a mathematical context.
    - o *p* and *q* often denote prime numbers or relatively prime numbers, or, in statistics, probabilities.
- *r* often denotes a remainder or modulus.
- *r*, *s*, and *t* usually play similar roles or are made to represent parallel notions in a mathematical context.

- u and v usually play similar roles or are made to represent parallel notions in a mathematical context, such as denoting a vertex (graph theory).
- w, x, y, and z usually play similar roles or are made to represent parallel notions in a mathematical context, such as representing unknowns in an equation.
- x, y and z correspond to the three Cartesian axes. In many two-dimensional cases, y will be expressed in terms of x; if a third dimension is added, z is expressed in terms of x and y.
    - o z typically denotes a complex number, or, in statistics, a normal random variate.
- α, β, γ, θ and φ commonly denote angle measures.
- ϵ usually represents an arbitrarily small positive number.
    - o ϵ and ä commonly denote two small positives.
- λ is used for eigenvalues.
- σ often denotes a sum, or, in statistics, the standard deviation.

Applied Statistics

In statistics, variables refer to measurable attributes, as these typically vary over time or between individuals. Variables can be discrete (taking values from a finite or countable set), continuous (having a continuous distribution function), or neither. Temperature is a continuous variable, while the number of legs of an animal is a discrete variable. This concept of a variable is widely used in the natural, medical, and social sciences.

In causal models, a distinction is made between "independent variables" and "dependent variables", the latter being expected to vary in value in response to changes in the former. In other words, an independent variable is presumed to potentially affect a dependent one. In experiments, independent variables include factors that can be altered or chosen by the researcher independent of other factors.

So, in an experiment to test if the boiling point of water changes with altitude, the altitude is under direct control and is the *independent* variable, and the boiling point is presumed to depend upon it, so being the *dependent* variable. The results of an experiment, or information to be used to draw conclusions, are known as data. It is often important to consider which variables to allow, or directly control or eliminate, in the design of experiments.

There are also quasi-independent variables, which are used by researchers to group things without affecting the variable itself. For example, to separate people into groups by their sex does not change whether they are male or female. Or a researcher may separate people, arbitrarily, on the amount of coffee they had drunk before beginning an experiment. The researcher cannot change the past, but can use it to split people into groups.

While independent variables can refer to quantities and qualities that are under experimental control, they can also include extraneous factors that influence results in a confusing or undesired manner. In statistics the technique to work this out is called correlation. If strongly confounding variables exist that can substantially change the result, it makes it harder to interpret. For example, a study on cancer against age will also have to take into account *variables such* as income, location, stress, and lifestyle. Without considering these, the results could be grossly inaccurate deductions. Because of this, controlling unwanted variables is important in research.

Number

A number is a mathematical object used to count and measure. In mathematics, the definition of number has been extended over the years to include such numbers as zero, negative numbers, rational numbers, irrational numbers, and complex numbers. Mathematical operations are certain procedures that take one or more numbers as input and produce a number as output. Unary operations take a single input number and produce a single output number.

For example, the successor operation adds one to an integer, thus the successor of 4 is 5. Binary operations take two input numbers and produce a single output number. Examples of binary operations include addition, subtraction, multiplication, division, and exponentiation. The study of numerical operations is called arithmetic. A notational symbol that represents a number is called a numeral. In addition to their use in counting and measuring, numerals are often used for labels (telephone numbers), for ordering (serial numbers), and for codes (e.g., ISBNs).

In common use, the word *number* can mean the abstract object, the symbol, or the word for the number.

Classification of Numbers

Different types of numbers are used in many cases. Numbers can be classified into sets, called number systems.

	Number Systems	
ℕ	Natural	0, 1, 2, 3, 4, ... or 1, 2, 3, 4, ...
ℤ	Integers	..., -5, -4, -3, -2, -1, 0, 1, 2, 3, 4, 5, ...
-	Positive integers	1, 2, 3, 4, 5, ...
ℚ	Rational	$^{a}/_{b}$ where a and b are integers and b is not zero
ℝ	Real	The limit of a convergent sequence of rational numbers
ℂ	Complex	$a + bi$ where a and b are real numbers and i is the square root of -1

Natural Numbers

The most familiar numbers are the natural numbers or counting numbers: one, two, three, and so on. Traditionally, the sequence of natural numbers started with 1 (0 was not even considered a number for the Ancient Greeks.) However, in the 19th century, set theorists and other mathematicians started including 0 (cardinality of the empty set, i.e. 0 elements, where 0 is thus the smallest cardinal number) in the set of natural numbers. Today, different mathematicians use the term to describe both sets, including zero or not. The mathematical symbol for the set of all natural numbers is N, also written $\mathbb{N}$.

In the base ten numeral system, in almost universal use today for mathematical operations, the symbols for natural numbers are written using ten digits: 0, 1, 2, 3, 4, 5, 6, 7, 8, and 9. In this base ten system, the rightmost digit of a natural number has a place value of one, and every other digit has a place value ten times that of the place value of the digit to its right.

In set theory, which is capable of acting as an axiomatic foundation for modern mathematics, natural numbers can be represented by classes of equivalent sets. For instance, the number 3 can be represented as the class of all sets that have exactly three elements. Alternatively, in Peano Arithmetic, the number 3 is represented as sss0, where s is the "successor" function (i.e., 3 is the third successor of 0). Many different representations are possible; all that is needed to formally represent 3 is to inscribe a certain symbol or pattern of symbols three times.

Integers

The negative of a positive integer is defined as a number that produces zero when it is added to the corresponding positive integer.

Negative numbers are usually written with a negative sign (a minus sign). As an example, the negative of 7 is written -7, and 7 + (-7) = 0. When the set of negative numbers is combined with the set of natural numbers (which includes zero), the result is defined as the set of integer numbers, also called integers, Z also written $\mathbb{Z}$. Here the letter Z comes from German *Zahl*, meaning "number".

The set of integers forms a ring with operations addition and multiplication.

Rational Numbers

A rational number is a number that can be expressed as a fraction with an integer numerator and a non-zero natural number denominator. Fractions are written as two numbers, the numerator and the denominator, with a dividing bar between them. In the fraction written $^m/_n$ or

$$\frac{m}{n}$$

m represents equal parts, where *n* equal parts of that size make up one whole. Two different fractions may correspond to the same rational number; for example $^1/_2$ and $^2/_4$ are equal, that is:

$$\frac{1}{2}=\frac{2}{4}.$$

If the absolute value of *m* is greater than *n*, then the absolute value of the fraction is greater than 1. Fractions can be greater than, less than, or equal to 1 and can also be positive, negative, or zero. The set of all rational numbers includes the integers, since every integer can be written as a fraction with denominator 1. For example -7 can be written $-^7/_1$. The symbol for the rational numbers is Q (for quotient), also written $\mathbb{Q}$.

Real Numbers

The real numbers include all of the measuring numbers. Real numbers are usually written using decimal numerals, in which a decimal point is placed to the right of the digit with place value one. Each digit to the right of the decimal point has a place value one-tenth of the place value of the digit to its left. Thus : 123.456 represents 1 hundred, 2 tens, 3 ones, 4 tenths, 5 hundredths, and 6 thousandths. In saying the number, the decimal is read "point", thus: "one two three point four five six ". In the US and UK and a number of other countries, the decimal point is represented by a period, whereas in continental Europe and certain other countries the decimal point is represented by a comma. Zero is often written as 0.0 when it must be treated as a real number rather than an integer. In the US and UK a number between "1 and 1 is always written with a leading zero to emphasize the decimal. Negative real numbers are written with a preceding minus sign:-123.456.

Every rational number is also a real number. It is not the case, however, that every real number is rational. If a real number cannot be written as a fraction of two integers, it is called irrational. A decimal that can be written as a fraction either ends (terminates) or forever repeats, because it is the answer to a problem in division. Thus the real number 0.5 can be written as $^1/_2$ and the real number 0.333... (forever repeating threes, otherwise written 0.3) can be written as $^1/_3$. On the other hand, the real number φ (pi), the ratio of the circumference of any circle to its diametre, is

$$\pi = 3.14159265358979....$$

Since the decimal neither ends nor forever repeats, it cannot be written as a fraction, and is an example of an irrational number. Other irrational numbers include

$$\sqrt{2} = 1.41421356237...$$

(the square root of 2, that is, the positive number whose square is 2).

Thus 1.0 and 0.999... are two different decimal numerals representing the natural number 1. There are infinitely many other ways of representing the number 1, for example $^2/_2$, $^3/_3$, 1.00, 1.000, and so on.

Every real number is either rational or irrational. Every real number corresponds to a point on the number line. The real numbers also have an important but highly technical property called the least upper bound property. The symbol for the real numbers is R, also written as $\mathbb{R}$.

When a real number represents a measurement, there is always a margin of error. This is often indicated by rounding or truncating a decimal, so that digits that suggest a greater accuracy than the measurement itself are removed. The remaining digits are called significant digits. For example, measurements with a ruler can seldom be made without a margin of error of at least 0.001 metres. If the sides of a rectangle are measured as 1.23 metres and 4.56 metres, then multiplication gives an area for the rectangle of 5.6088 square metres. Since only the first two digits after the decimal place are significant, this is usually rounded to 5.61. In abstract algebra, it can be shown that any complete ordered field is isomorphic to the real numbers. The real numbers are not, however, an algebraically closed field.

Complex Numbers

Moving to a greater level of abstraction, the real numbers can be extended to the complex numbers. This set of numbers arose, historically, from trying to find closed formulas for the roots of cubic and quartic polynomials. This led to expressions involving the square roots of negative numbers, and eventually to the definition of a new number: the square root of negative one, denoted by *i*, a symbol assigned by Leonhard Euler, and called the imaginary unit. The complex numbers consist of all numbers of the form

$$a + bi$$

where *a* and *b* are real numbers. In the expression $a + bi$, the real number *a* is called the real part and *b* is called the imaginary part. If the real part of a complex number is zero, then the number is called an imaginary number or is referred to as *purely imaginary*; if the imaginary part is zero, then the number is a real number. Thus the real numbers are a subset of the complex numbers. If the real and imaginary parts of a complex number are both integers, then the number is called a Gaussian integer. The symbol for the complex numbers is C or $\mathbb{C}$.

In abstract algebra, the complex numbers are an example of an algebraically closed field, meaning that every polynomial with complex coefficients can be factored into linear factors. Like the real number system, the complex number system is a field and is complete, but unlike the real numbers it is not ordered. That is, there is no meaning in saying that *i* is greater than 1, nor is there any meaning in saying that *i* is less than 1. In technical terms, the complex numbers lack the trichotomy property.

Complex numbers correspond to points on the complex plane, sometimes called the Arg and plane.

Each of the number systems mentioned above is a proper subset of the next number system. Symbolically, $\mathbb{N} \subset \mathbb{Z} \subset \mathbb{Q} \subset \mathbb{R} \subset \mathbb{C}$.

Computable Numbers

Moving to problems of computation, the computable numbers are determined in the set of the real numbers. The computable numbers, also known as the recursive numbers or the computable reals, are the real numbers that can be computed to within any desired precision by a finite, terminating algorithm. Equivalent definitions can be given using μ-recursive functions, Turing machines or λ-calculus as the formal representation of algorithms. The computable numbers form a real closed field and can be used in the place of real numbers for many, but not all, mathematical purposes.

Other Types

Algebraic numbers are those that can be expressed as the solution to a polynomial equation with integer coefficients. The complement of the algebraic numbers are the transcendental numbers.

Hyperreal numbers are used in non-standard analysis. The hyperreals, or nonstandard reals (usually denoted as *R), denote an ordered field that is a proper extension of the ordered field of real numbers R and satisfies the transfer principle. This principle allows true first order statements about R to be reinterpreted as true first order statements about *R.

Superreal and surreal numbers extend the real numbers by adding infinitesimally small numbers and infinitely large numbers, but still form fields.

The p-adic numbers may have infinitely long expansions to the left of the decimal point, in the same way that real numbers may have infinitely long expansions to the right. The number system that results depends on what base is used for the digits: any base is possible, but a prime number base provides the best mathematical properties.

For dealing with infinite collections, the natural numbers have been generalised to the ordinal numbers and to the cardinal numbers. The former gives the ordering of the collection, while the latter gives its size. For the finite set, the ordinal and cardinal numbers are equivalent, but they differ in the infinite case.

A relation number is defined as the class of relations consisting of all those relations that are similar to one member of the class.

Sets of numbers that are not subsets of the complex numbers are sometimes called hyper complex numbers. They include the quaternions H, invented by Sir William Rowan Hamilton, in which multiplication is not commutative, and the octonions, in which multiplication is not associative. Elements of function fields of non-zero characteristic behave in some ways like numbers and are often regarded as numbers by number theorists.

Specific Uses

There are also other sets of numbers with specialised uses. Some are subsets of the complex numbers. For example, algebraic numbers are the roots of polynomials with rational coefficients. Complex numbers that are not algebraic are called transcendental numbers.

An even number is an integer that is "evenly divisible" by 2, i.e., divisible by 2 without remainder; an odd number is an integer that is not evenly divisible by 2. (The old-fashioned term "evenly divisible" is now almost always shortened to "divisible".) A formal definition of an odd number is that it is an integer of the form $n = 2k + 1$, where k is an integer. An even number has the form $n = 2k$ where k is an integer.

A perfect number is a positive integer that is the sum of its proper positive divisors—the sum of the positive divisors not including the number itself. Equivalently, a perfect number is a number that is half the sum of all of its positive divisors, or $\sigma(n) = 2\,n$. The first perfect number is 6, because 1, 2, and 3 are its proper positive divisors and $1 + 2 + 3 = 6$. The next perfect number is $28 = 1 + 2 + 4 + 7 + 14$. The next perfect numbers are 496 and 8128 (sequence A000396 in OEIS). These first four perfect numbers were the only ones known to early Greek mathematics.

A figurate number is a number that can be represented as a regular and discrete geometric pattern (e.g. dots). If the pattern is polytopic, the figurate is labelled a polytopic number, and may be a polygonal number or a polyhedral number. Polytopic numbers for r = 2, 3, and 4 are:

- $P_2(\mathrm{n}) = {}^1\mathrm{D}_{\,2}\; n(n + 1)$ (triangular numbers)
- $P_3(\mathrm{n}) = {}^1\mathrm{D}_{\,6}\; n(n + 1)(n + 2)$ (tetrahedral numbers)
- $P_4(\mathrm{n}) = {}^1\mathrm{D}_{\,24}\; n(n + 1)(n + 2)(n + 3)$ (pentatopic numbers).

Numerals

Numbers should be distinguished from numerals, the symbols used to represent numbers. Boyer showed that Egyptians created the first ciphered numeral system. Greeks followed by mapping their counting numbers onto Ionian and Doric alphabets. The number five can be represented by both the base ten numeral '5', by the Roman numeral 'd!' and ciphered letters. Notations used to represent numbers are discussed in the article numeral systems. An important development in the history of numerals was the development of a positional system, like modern decimals, which can represent very large numbers. The Roman numerals require extra symbols for larger numbers.

History

The First Use of Numbers: Bones and other artifacts have been discovered with marks cut into them which many believe are tally marks. These tally marks may have been used for counting elapsed time, such as numbers of days, lunar cycles or keeping records of quantities, such as of animals.

A tallying system has no concept of place value (as in modern decimal notation), which limits its representation of large numbers. Nonetheless tallying systems are considered the first kind of abstract numeral system.

The first known system with place value was the Mesopotamian base 60 system (ca. 3400 BC) and the earliest known base 10 system dates to 3100 BC in Egypt.

Zero

The use of zero as a number should be distinguished from its use as a placeholder numeral in place-value systems. Many ancient texts used zero. Babylonian and Egyptian texts used it. Egyptians used the word *nfr* to denote zero balance in double entry accounting entries. Indian texts used a Sanskrit word *Shunya* to refer to the concept of *void*. In mathematics texts this word often refers to the number zero.

Records show that the Ancient Greeks seemed unsure about the status of zero as a number: they asked themselves "how can 'nothing' be something?" leading to interesting philosophical and, by the Medieval period, religious arguments about the nature and existence of zero and the vacuum. The paradoxes of Zeno of Elea depend in large part on the uncertain interpretation of zero. (The ancient Greeks even questioned whether 1 was a number.)

The late Olmec people of south-central Mexico began to use a true zero (a shell glyph) in the New World possibly by the 4th century BC but certainly by 40 BC, which became an integral part of Maya numerals and the Maya calendar. Mayan arithmetic used base 4 and base 5 written as base 20. Sanchez in 1961 reported a base 4, base 5 'finger' abacus.

By 130 AD, Ptolemy, influenced by Hipparchus and the Babylonians, was using a symbol for zero (a small circle with a long overbar) within a sexagesimal numeral system otherwise using alphabetic Greek numerals. Because it was used alone, not as just a placeholder, this Hellenistic zero was the first *documented* use of a true zero in the Old World. In later Byzantine manuscripts of his *Syntaxis Mathematica* (*Almagest*), the Hellenistic zero had morphed into the Greek letter omicron (otherwise meaning 70).

Another true zero was used in tables alongside Roman numerals by 525 (first known use by Dionysius Exiguus), but as a word, *nulla* meaning *nothing*, not as a symbol. When division produced zero as a remainder, *nihil*, also meaning *nothing*, was used. These medieval zeros were used by all future medieval computists (calculators of Easter). An isolated use of their initial, N, was used in a table of Roman numerals by Bede or a colleague about 725, a true zero symbol.

An early documented use of the zero by Brahmagupta (in the Brahmasphuta siddhanta) dates to 628. He treated zero as a number and discussed operations involving it, including division. By this time (the 7th century) the concept had clearly reached Cambodia as Khmer numerals, and documentation shows the idea later spreading to China and the Islamic world.

Negative Numbers

The abstract concept of negative numbers was recognised as early as 100 BC – 50 BC. The Chinese Nine *Chapters on the Mathematical Art* (Chinese: Jiu-zhang Suanshu) contains methods for finding the areas of figures; red rods were used to denote positive coefficients, black for negative. This is the earliest known mention of negative numbers in the East; the first reference in a Western work was in the 3rd century in Greece. Diophantus referred to the equation equivalent to $4x + 20 = 0$ (the solution is negative) in *Arithmetica*, saying that the equation gave an absurd result.

During the 600s, negative numbers were in use in India to represent debts. Diophantus' previous reference was discussed more explicitly by Indian mathematician Brahmagupta, in Brahma-Sphuta-Siddhanta 628, who used negative numbers to produce the general form quadratic formula that remains in use today. However, in the 12th century in India, Bhaskara gives negative roots for quadratic equations but says the negative value "is in this case not to be taken, for it is inadequate; people do not approve of negative roots."

European mathematicians, for the most part, resisted the concept of negative numbers until the 17th century, although Fibonacci allowed negative solutions in financial problems where they could be interpreted as debts (chapter 13 of *Liber Abaci*, 1202) and later as losses (in *Flos*).

At the same time, the Chinese were indicating negative numbers either by drawing a diagonal stroke through the right-most nonzero digit of the corresponding positive number's numeral. The first use of negative numbers in a European work was by Chuquet during the 15th century. He used them as exponents, but referred to them as "absurd numbers".

As recently as the 18th century, it was common practice to ignore any negative results returned by equations on the assumption that they were meaningless, just as René Descartes did with negative solutions in a Cartesian coordinate system.

Rational Numbers

It is likely that the concept of fractional numbers dates to prehistoric times. The Ancient Egyptians used their Egyptian fraction notation for rational numbers in mathematical texts such as the Rhind Mathematical Papyrus and the Kahun Papyrus. Classical Greek and Indian mathematicians made studies of the theory of rational numbers, as part of the general study of number theory. The best known of these is Euclid's Elements, dating to roughly 300 BC. Of the Indian texts, the most relevant is the Sthananga Sutra, which also covers number theory as part of a general study of mathematics.

The concept of decimal fractions is closely linked with decimal place-value notation; the two seem to have developed in tandem. For example, it is common for the Jain math sutras to include calculations of decimal-fraction approximations to pi or the square root of two. Similarly, Babylonian math texts had always used sexagesimal (base 60) fractions with great frequency.

Irrational Numbers

The earliest known use of irrational numbers was in the Indian Sulba Sutras composed between 800–500 BC. The first existence proofs of irrational numbers is usually attributed to Pythagoras, more specifically to the Pythagorean Hippasus of Metapontum, who produced a (most likely geometrical) proof of the irrationality of the square root of 2. The story goes that Hippasus discovered irrational numbers when trying to represent the square root of 2 as a fraction. However Pythagoras believed in the absoluteness of numbers, and could not accept the existence of irrational numbers. He could not disprove their existence through logic, but he could not accept irrational numbers, so he sentenced Hippasus to death by drowning. The sixteenth century brought final European acceptance of negative integral and fractional numbers. By the seventeenth century, mathematicians generally used decimal fractions with modern notation. It was not, however, until the nineteenth century that mathematicians separated irrationals into algebraic and transcendental parts, and once more undertook scientific study of irrationals.

It had remained almost dormant since Euclid. 1872 brought publication of the theories of Karl Weierstrass (by his pupil Kossak), Heine (*Crelle*, 74), Georg Cantor (Annalen, 5), and Richard Dedekind. In 1869, Méray had taken the same point of departure as Heine, but the theory is generally referred to the year 1872. Weierstrass's method was completely set forth by Salvatore Pincherle (1880), and Dedekind's has received additional prominence through the author's later work (1888) and endorsement by Paul Tannery (1894).

Weierstrass, Cantor, and Heine base their theories on infinite series, while Dedekind founds his on the idea of a cut (Schnitt) in the system of real numbers, separating all rational numbers into two groups having certain characteristic properties. The subject has received later contributions at the hands of Weierstrass, Kronecker (Crelle, 101), and Méray.

Continued fractions, closely related to irrational numbers (and due to Cataldi, 1613), received attention at the hands of Euler, and at the opening of the nineteenth century were brought into prominence through the writings of Joseph Louis Lagrange. Other noteworthy contributions have been made by Drucken müller (1837), Kunze (1857), Lemke (1870), and Günther (1872). Ramus (1855) first connected the subject with determinants, resulting, with the subsequent contributions of Heine, Mφbius, and Günther, in the theory of Ketten bruch determinanten. Dirichlet also added to the general theory, as have numerous contributors to the applications of the subject.

Transcendental Numbers and Reals

The first results concerning transcendental numbers were Lambert's 1761 proof that φ cannot be rational, and also that e^n is irrational if n is rational (unless $n = 0$). (The constant e was first referred to in Napier's 1618 work on logarithms.) Legendre extended this proof to show that φ is not the square root of a rational number. The search for roots of quintic and higher degree equations was an important development, the Abel–Ruffini theorem (Ruffini 1799, Abel 1824) showed that they could not be solved by radicals (formula involving only arithmetical operations and roots). Hence it was necessary to consider the wider set of algebraic numbers (all solutions to polynomial equations). Galois (1832) linked polynomial equations to group theory giving rise to the field of Galois theory.

The existence of transcendental numbers was first established by Liouville (1844, 1851). Hermite proved in 1873 that *e* is transcendental and Lindemann proved in 1882 that φ is transcendental. Finally Cantor shows that the set of all real numbers is uncountably infinite but the set of all algebraic numbers is countably infinite, so there is an uncountably infinite number of transcendental numbers.

Infinity and Infinitesimals

The earliest known conception of mathematical infinity appears in the Yajur Veda, an ancient Indian script, which at one point states, "If you remove a part from infinity or add a part to infinity, still what remains is infinity." Infinity was a popular topic of philosophical study among the Jain mathematicians c. 400 BC. They distinguished between five types of infinity: infinite in one and two directions, infinite in area, infinite everywhere, and infinite perpetually.

Aristotle defined the traditional Western notion of mathematical infinity. He distinguished between actual infinity and potential infinity—the general consensus being that only the latter had true value. Galileo's Two New Sciences discussed the idea of one-to-one correspondences between infinite sets. But the next major advance in the theory was made by Georg Cantor; in 1895 he published a book about his new set theory, introducing, among other things, transfinite numbers and formulating the continuum hypothesis. This was the first mathematical model that represented infinity by numbers and gave rules for operating with these infinite numbers.

In the 1960s, Abraham Robinson showed how infinitely large and infinitesimal numbers can be rigorously defined and used to develop the field of nonstandard analysis. The system of hyperreal numbers represents a rigorous method of treating the ideas about infinite and infinitesimal numbers that had been used casually by mathematicians, scientists, and engineers ever since the invention of infinitesimal calculus by Newton and Leibniz.

A modern geometrical version of infinity is given by projective geometry, which introduces "ideal points at infinity," one for each spatial direction. Each family of parallel lines in a given direction is postulated to converge to the corresponding ideal point. This is closely related to the idea of vanishing points in perspective drawing.

Complex Numbers

The earliest fleeting reference to square roots of negative numbers occurred in the work of the mathematician and inventor Heron of Alexandria in the 1st century AD, when he considered the volume of an impossible frustum of a pyramid. They became more prominent when in the 16th century closed formulas for the roots of third and fourth degree polynomials were discovered by Italian mathematicians such as Niccolo Fontana Tartaglia and Gerolamo Cardano. It was soon realised that these formulas, even if one was only interested in real solutions, sometimes required the manipulation of square roots of negative numbers.

This was doubly unsettling since they did not even consider negative numbers to be on firm ground at the time. When René Descartes coined the term "imaginary" for these quantities in 1637, he intended it as derogatory. A further source of confusion was that the equation

$$\left(\sqrt{-1}\right)^2 = \sqrt{-1}\sqrt{-1} = -1$$

seemed to be capriciously inconsistent with the algebraic identity

$$\sqrt{a}\sqrt{b} = \sqrt{ab},$$

which is valid for positive real numbers a and b, and was also used in complex number calculations with one of a, b positive and the other negative. The incorrect use of this identity, and the related identity

$$\frac{1}{\sqrt{a}} = \sqrt{\frac{1}{a}}$$

in the case when both a and b are negative even bedeviled Euler. This difficulty eventually led him to the convention of using the special symbol i in place of $\sqrt{-1}$ to guard against this mistake.

The 18th century saw the work of Abraham de Moivre and Leonhard Euler. de Moivre's formula (1730) states:

$$(\cos\theta + i\sin\theta)^n = \cos n\theta + i\sin n\theta$$

and to Euler (1748) Euler's formula of complex analysis:

$$\cos\theta + i\sin\theta = e^{i\theta}.$$

The existence of complex numbers was not completely accepted until Caspar Wessel described the geometrical interpretation in 1799. Carl Friedrich Gauss rediscovered and popularised it several years later, and as a result the theory of complex numbers received a notable expansion. The idea of the graphic representation of complex numbers had appeared, however, as early as 1685, in Wallis's *De Algebra tractatus*.

Also in 1799, Gauss provided the first generally accepted proof of the fundamental theorem of algebra, showing that every polynomial over the complex numbers has a full set of solutions in that realm. The general acceptance of the theory of complex numbers is due to the labours of Augustin Louis Cauchy and Niels Henrik Abel, and especially the latter, who was the first to boldly use complex numbers with a success that is well known.

Gauss studied complex numbers of the form $a + bi$, where a and b are integral, or rational (and i is one of the two roots of $x^2 + 1 = 0$). His student, Gotthold Eisenstein, studied the type $a + b\omega$, where ω is a complex root of $x^3 - 1 = 0$. Other such classes (called cyclotomic fields) of complex numbers derive from the roots of unity $x^k - 1 = 0$ for higher values of k. This generalisation is largely due to Ernst Kummer, who also invented ideal numbers, which were expressed as geometrical entities by Felix Klein in 1893. The general theory of fields was created by Évariste Galois, who studied the fields generated by the roots of any polynomial equation $F(x) = 0$.

In 1850 Victor Alexandre Puiseux took the key step of distinguishing between poles and branch points, and introduced the concept of essential singular points. This eventually led to the concept of the extended complex plane.

Prime Numbers

Prime numbers have been studied throughout recorded history. Euclid devoted one book of the *Elements* to the theory of primes; in it he proved the infinitude of the primes and the fundamental theorem of arithmetic, and presented the Euclidean algorithm for finding the greatest common divisor of two numbers.

In 240 BC, Eratosthenes used the Sieve of Eratosthenes to quickly isolate prime numbers. But most further development of the theory of primes in Europe dates to the Renaissance and later eras.

In 1796, Adrien-Marie Legendre conjectured the prime number theorem, describing the asymptotic distribution of primes. Other results concerning the distribution of the primes include Euler's proof that the sum of the reciprocals of the primes diverges, and the Goldbach conjecture, which claims that any sufficiently large even number is the sum of two primes. Yet another conjecture related to the distribution of prime numbers is the Riemann hypothesis, formulated by Bernhard Riemann in 1859. The prime number theorem was finally proved by Jacques Hadamard and Charles de la Vallée-Poussinin 1896. Goldbach and Riemann's conjectures remain unproven and unrefuted.

Word alternatives

Some numbers traditionally have alternative words to express them, including the following:

- Pair, couple, brace: 2
- Trio: 3
- Dozen: 12
- Baker's dozen: 13
- Score: 20
- Gross: 144
- Ream: 480 (old measure) 500 (new measure)
- Great gross: 1728
- "-figure", as in digit, generally for larger-number ranges
- "five-figure": 10,000 to 99,999 (five digits)
- "six-figure": 100,000 to 999,999 (six digits)

... *et cetera.*

Operation

The general operation as explained on this page should not be confused with the more specificoperators on vector spaces.

In its simplest meaning in mathematics and logic, an operation is an action or procedure which produces a new value from one or more input values. There are two common types of operations: unary and binary. Unary operations involve only one value, such as negation and trigonometric functions. Binary operations, on the other hand, take two values, and include addition, subtraction, multiplication, division, and exponentiation.

Operations can involve mathematical objects other than numbers. The logical values *true* and *false* can be combined using logic operations, such as *and*, *or,* and *not.* Vectors can be added and subtracted. Rotations can be combined using the function composition operation, performing the first rotation and then the second. Operations on sets include the binary operations *union* and *intersection* and the unary operation of *complementation.*

Operations on functions include composition and convolution. Operations may not be defined for every possible value. For example, in the real numbers one cannot divide by zero or take square roots of negative numbers. The values for which an operation is defined form a set called its *domain*. The set which contains the values produced is called the *codomain*, but the set of actual values attained by the operation is its *range*. For example, in the real numbers, the squaring operation only produces nonnegative numbers; the codomain is the set of real numbers but the range is the nonnegative numbers.

Operations can involve dissimilar objects. A vector can be multiplied by a scalar to form another vector. And the inner product operation on two vectors produces a scalar. An operation may or may not have certain properties, for example it may be associative, commutative, anticommutative, idempotent, and so on.

The values combined are called *operands*, *arguments*, or *inputs*, and the value produced is called the value, *result*, or *output*. Operations can have fewer or more than two inputs.

An operation is like an operator, but the point of view is different. For instance, one often speaks of "the operation of addition" or "addition operation" when focusing on the operands and result, but one says "addition operator" (rarely "operator of addition") when focusing on the process, or from the more abstract viewpoint, the function $+: S \times S \rightarrow S$.

General Definition

An operation ω *is a function of the form* $\omega : V \rightarrow Y$, *where* $V \subset X_1 \times \ldots \times X_k$. *The sets* X_k *are called the* domains *of the operation, the set* Y *is called the* codomain *of the operation, and the fixed non-negative integer* k *(the number of arguments) is called the* type *or* arity *of the operation. Thus a unary operation has arity one, and a binary operation has arity two. An operation of arity zero, called a* nullary *operation, is simply an element of the codomain* Y. *An operation of arity* k *is called a* k*-ary operation. Thus a* k*-ary operation is a* (k+1)*-ary relation that is functional on its first* k *domains.*

The above describes what is usually called a finitary *operation, referring to the finite number of arguments (the value* k*). There are obvious extensions where the arity is taken to be an infinite ordinal or cardinal, or even an arbitrary set indexing the arguments.*

Often, use of the term operation *implies that the domain of the function is a power of the codomain (i.e. the Cartesian product of one or more copies of the codomain), although this is by no means universal, as in the example of multiplying a vector by a scalar.*

Thus, since k *can be 1, in the most general sense given here,* operation *is synonymous with function, map and mapping, that is, a relation, for which each element of the domain (input set) is associated with exactly one element of the codomain (set of possible outputs).*

Binary Relation

In mathematics, a binary relation on a set A is a collection of ordered pairs of elements of A. In other words, it is a subset of the Cartesian product $A^2 = A \times A$. More generally, a binary relation between two sets A and B is a subset of $A \times B$. The terms dyadic relation and 2-place relation are synonyms for binary relations.

An example is the "divides" relation between the set of prime numbers P and the set of integers Z, in which every prime p is associated with every integer z that is a multiple of p (and not with any integer that is not a multiple of p). In this relation, for instance, the prime 2 is associated with numbers that include "4, 0, 6, 10, but not 1 or 9; and the prime 3 is associated with numbers that include 0, 6, and 9, but not 4 or 13.

Binary relations are used in many branches of mathematics to model concepts like "is greater than", "is equal to", and "divides" in arithmetic, "is congruent to" in geometry, "is adjacent to" in graph theory, "is orthogonal to" in linear algebra and many more. The concept of function is defined as a special kind of binary relation. Binary relations are also heavily used in computer science.

A binary relation is the special case $n = 2$ of an n-ary relation $R \subseteq A_1 \times \ldots \times A_n$, that is, a set of n-tuples where the jth component of each n-tuple is taken from the jth domain A_j of the relation.

In some systems of axiomatic set theory, relations are extended to classes, which are generalisations of sets. This extension is needed for, among other things, modelling the concepts of "is an element of" or "is a subset of" in set theory, without running into logical inconsistencies such as Russell's paradox.

Formal Definition

A binary relation R *is usually defined as an ordered triple (*X, Y, G*) where* X *and* Y *are arbitrary sets (or classes), and* G *is a subset of the Cartesian product* X × Y. *The sets* X *and* Y *are called the domain (or* *the set of departure) and codomain (or the set of destination), respectively, of the relation, and* G *is called its graph.*

The statement (x,y) " R *is read "*x *is* R*-related to* y*", and is denoted by* xRy *or* R(x,y)*. The latter notation corresponds to viewing* R *as the characteristic function on "X" x "Y" for the set of pairs of* G.

The order of the elements in each pair of G *is important: if* a $\neq$ b, *then* aRb *and* bRa *can be true or false, independently of each other.*

Is a Relation More than its Graph?

According to the definition above, two relations with the same graph may be different, if they differ in the sets X and Y. For example, if $G = \{(1,2),(1,3),(2,7)\}$, then (Z,Z, G), (R, N, G), and (N, R, G) are three distinct relations.

Some mathematicians do not consider the sets X and Y to be part of the relation, and therefore define a binary relation as being a subset of $X \times Y$, that is, just the graph G. According to this view, the set of pairs $\{(1,2),(1,3),(2,7)\}$ is a relation from any set that contains $\{1,2\}$ to any set that contains $\{2,3,7\}$.

A special case of this difference in points of view applies to the notion of function. Most authors insist on distinguishing between a function's codomain and its range. Thus, a single "rule," like mapping every real number x to x^2, can lead to distinct functions $f{:}R \to R$ and $g{:}R \to R^+$, depending on whether the images under that rule are understood to be reals or, more restrictively, non-negative reals. But others view functions as simply sets of ordered pairs with unique first components. This difference in perspectives does raise some nontrivial issues. As an example, the former camp considers surjectivity—or being onto—as a property of functions, while the latter sees it as a relationship that functions may bear to sets.

Either approach is adequate for most uses, provided that one attends to the necessary changes in language, notation, and the definitions of concepts like restrictions, composition, inverse relation, and so on. The choice between the two definitions usually matters only in very formal contexts, like category theory.

Example

Example: Suppose there are four objects {ball, car, doll, gun} and four persons {John, Mary, Ian, Venus}. Suppose that John owns the ball, Mary owns the doll, and Venus owns the car. Nobody owns the gun and Ian owns nothing. Then the binary relation "is owned by" is given as

R=({ball, car, doll, gun}, {John, Mary, Ian, Venus}, {(ball, John), (doll, Mary), (car, Venus)}).

Thus the first element of R is the set of objects, the second is the set of people, and the last element is a set of ordered pairs of the form (object, owner).

The pair (ball, John), denoted by ${}_{\text{ball}}R_{\text{John}}$ means that the ball is owned by John.

Two different relations could have the same graph. For example: the relation

({ball, car, doll, gun}, {John, Mary, Venus}, {(ball,John), (doll, Mary), (car, Venus)})

is different from the previous one as everyone is an owner. But the graphs of the two relations are the same.

Nevertheless, *R* is usually identified or even defined as G(*R*) and "an ordered pair (*x*, *y*) " G(*R*)" is usually denoted as "(*x*, *y*) "*R*".

Special Types of Binary Relations

Some important classes of binary relations R *between* X *and* Y *are listed below.*

Uniqueness properties:

- injective (also called left-unique): for all *x* and *z* in *X* and *y* in *Y* it holds that if *xRy* and *zRy* then *x* = *z*.
- functional (also called right-unique or right-definite): for all *x* in *X*, and *y* and *z* in *Y* it holds that if *xRy* and *xRz* then *y* = *z*; such a binary relation is called a partial function.
- one-to-one (also written 1-to-1): injective and functional.

Totality properties:

- left-total: for all *x* in *X* there exists a *y* in *Y* such that *xRy* (this property, although sometimes also referred to as *total*, is different from the definition of *total* in the next section).
- surjective (also called right-total): for all *y* in *Y* there exists an *x* in *X* such that *xRy*.
- A correspondence: a binary relation that is both left-total and surjective.

Uniqueness and totality properties:

- A function: a relation that is functional and left-total.
- A bijection: a one-to-one correspondence; such a relation is a function and is said to be bijective.

Relations Over a Set

If X = Y then we simply say that the binary relation is over X. Or it is an endorelation over X. Some classes of endorelations are widely studied in graph theory, where they're known as directed graphs.

The set of all binary relations B(X) on a set X is a semigroup with involution with the involution being the mapping of a relation to its inverse relation.

Some important classes of binary relations over a set X *are:*

- reflexive: for all x in X it holds that xRx. For example, "greater than or equal to" is a reflexive relation but "greater than" is not.
- irreflexive (or strict): for all x in X it holds that not xRx. "Greater than" is an example of an irreflexive relation.
- coreflexive: for all x and y in X it holds that if xRy then $x = y$. "Equal to and odd" is an example of a coreflexive relation.
- symmetric: for all x and y in X it holds that if xRy then yRx. "Is a blood relative of" is a symmetric relation, because x is a blood relative of y if and only if y is a blood relative of x.
- antisymmetric: for all *distinct* x and y in X, if xRy then not yRx.
- asymmetric: for all x and y in X, if xRy then not yRx. (So asymmetricity is stronger than anti-symmetry. In fact, asymmetry is equivalent to anti-symmetry plus irreflexivity.)
- transitive: for all x, y and z in X it holds that if xRy and yRz then xRz. (Note that, under the assumption of transitivity, irreflexivity and asymmetry are equivalent.)
- total: for all x and y in X it holds that xRy or yRx (or both). "Is greater than or equal to" is an example of a total relation (this definition for *total* is different from *left total* in the previous section).
- trichotomous: for all x and y in X exactly one of xRy, yRx or $x = y$ holds. "Is greater than" is an example of a trichotomous relation.
- Euclidean: for all x, y and z in X it holds that if xRy and xRz, then yRz (and zRy). Equality is a Euclidean relation because if x=y and x=z, then y=z.
- serial: for all x in X, there exists y in X such that xRy. "Is greater than" is a serial relation on the integers. But it is not a serial

relation on the positive integers, because there is no y in the positive integers such that $1>y$. However, the "Is less than" is a serial relation on the positive integers (the natural numbers), the rational numbers and the real numbers. Every reflexive relation is serial.

- set-like: for every x in X, the class of all y such that yRx is a set. (This makes sense only if we allow relations on proper classes.) The usual ordering $<$ on the class of ordinal numbers is set-like, while its inverse $>$ is not.

A relation that is reflexive, symmetric, and transitive is called an equivalence relation. A relation that is reflexive, antisymmetric, and transitive is called a partial order. A partial order that is total is called a total order, *simple order*, linear order, or a chain. A linear order where every nonempty set has a least element is called a well-order. A relation that is symmetric, transitive, and serial is also reflexive.

Operations on Binary Relations

If R is a binary relation over X and Y, then the following is a binary relation over Y and X:

- Inverse or converse: R^{-1}, defined as $R^{-1} = \{(y, x) \mid (x, y)\in\}$. A binary relation over a set is equal to its inverse if and only if it is symmetric.

If R is a binary relation over X, then each of the following is a binary relation over X:

- Reflexive closure: $R^{=}$, defined as $R^{=} = \{(x,x) \mid x \in X\} \cup R$ or the smallest reflexive relation over X containing R. This can be seen to be equal to the intersection of all reflexive relations containing R.
- Reflexive reduction: $R^{\neq}$, defined as $R^{\neq} = R \setminus \{(x,x) \mid x \in X\}$ or the largest irreflexive relation over X contained in R.
- Transitive closure: R^{+}, defined as the smallest transitive relation over X containing R. This can be seen to be equal to the intersection of all transitive relations containing R.
- Transitive reduction: R^{-}, defined as a minimal relation having the same transitive closure as R.
- Reflexive transitive closure: R^{*}, defined as $R^{*} = (R^{+})^{=}$, the smallest preorder containing R.

- Reflexive transitive symmetric closure: $R^{\equiv}$, defined as the smallest equivalence relation over X containing R.

If R, S are binary relations over X and Y, then each of the following is a binary relation:

- Union: $R \cup S \subseteq X \times Y$, defined as $R \cup S = \{(x, y) \mid (x, y) \in R \text{ or } (x, y) \in S\}$.
- Intersection: $R \cap S \subseteq X \times Y$, defined as $R \cap S = \{(x,y) \mid (x,y) \in R \text{ and } (x,y) \in S\}$.

If R is a binary relation over X and Y, and S is a binary relation over Y and Z, then the following is a binary relation over X and Z:

- Composition: $S \circ R$, also denoted $R\,;\,S$ (or more ambiguously $R \circ S$), defined as $S \circ R = \{ (x, z) \mid \text{there exists } y \in Y, \text{ such that } (x, y) \in R \text{ and } (y, z) \in S\}$. The order of R and S in the notation $S \circ R$, used here agrees with the standard notational order for composition of functions.

Complement

If R is a binary relation over X and Y, then the following too:

- The complement S is defined as $x\,S\,y$ if not $x\,R\,y$.

The complement of the inverse is the inverse of the complement.

If $X = Y$ the complement has the following properties:

- If a relation is symmetric, the complement is too.
- The complement of a reflexive relation is irreflexive and vice versa.
- The complement of a strict weak order is a total preorder and vice versa.

The complement of the inverse has these same properties.

Restriction

The restriction of a binary relation on a set X to a subset S is the set of all pairs (x, y) in the relation for which x and y are in S. If a relation is reflexive, irreflexive, symmetric, antisymmetric, asymmetric, transitive, total, trichotomous, a partial order, total order, strict weak order, total preorder (weak order), or an equivalence relation, its restrictions are too.

However, the transitive closure of a restriction is a subset of the restriction of the transitive closure, i.e., in general not equal. Also, the various concepts of completeness (not to be confused with being "total") do not carry over to restrictions. For example, on the set of real numbers a property of the relation "≤" is that every non-empty subset S of R with an upper bound in R has a least upper bound (also called supremum) in R. However, for a set of rational numbers this supremum is not necessarily rational, so the same property does not hold on the restriction of the relation "≤" to the set of rational numbers.

The *left-restriction* (*right-restriction*, respectively) of a binary relation between X and Y to a subset S of its domain (codomain) is the set of all pairs (x, y) in the relation for which x (y) is an element of S.

Sets versus Classes

Certain mathematical "relations", such as "equal to", "member of", and "subset of", cannot be understood to be binary relations as defined above, because their domains and codomains cannot be taken to be sets in the usual systems of axiomatic set theory. For example, if we try to model the general concept of "equality" as a binary relation =, we must take the domain and codomain to be the "set of all sets", which is not a set in the usual set theory. The usual work-around to this problem is to select a "large enough" set A, that contains all the objects of interest, and work with the restriction $=_A$ instead of =.

Similarly, the "subset of" relation ⊆ needs to be restricted to have domain and codomain P(A) (the power set of a specific setA): the resulting set relation can be denoted $\subseteq_A$. Also, the "member of" relation needs to be restricted to have domain A and codomain P(A) to obtain a binary relation $\in_A$ that is a set.

Another solution to this problem is to use a set theory with proper classes, such as NBG or Morse–Kelley set theory, and allow the domain and codomain (and so the graph) to be proper classes: in such a theory, equality, membership, and subset are binary relations without special comment. (A minor modification needs to be made to the concept of the ordered triple (X, Y, G), as normally a proper class cannot be a member of an ordered tuple; or of course one can identify the function with its graph in this context.)

In most mathematical contexts, references to the relations of equality, membership and subset are harmless because they can be understood implicitly to be restricted to some set in the context.

The Number of Binary Relations

The number of distinct binary relations on an n*-element set is* 2^{n^2} *(sequence A002416 in OEIS):*

Number of n-element binary relations of different types								
n	***all***	***transitive***	***reflexive***	***preorder***	***partial order***	***total preorder***	***total order***	***equivalence relation***
0	1	1	1	1	1	1	1	1
1	2	2	1	1	1	1	1	1
2	16	13	4	4	3	3	2	2
3	512	171	64	29	19	13	6	5
4	65536	3994	4096	355	219	75	24	15
OEIS	A002416	A006905	A053763	A000798	A001035	A000670	A000142	A000110

Notes:

- The number of irreflexive relations is the same as that of reflexive relations.
- The number of strict partial orders (irreflexive transitive relations) is the same as that of partial orders.
- The number of strict weak orders is the same as that of total preorders.
- The total orders are the partial orders that are also total preorders. The number of preorders that are neither a partial order nor a total preorder is, therefore, the number of preorders, minus the number of partial orders, minus the number of total preorders, plus the number of total orders: 0, 0, 0, 3, and 85, respectively.
- the number of equivalence relations is the number of partitions, which is the Bell number.

The binary relations can be grouped into pairs (relation, complement), except that for $n = 0$ the relation is its own complement. The non-symmetric ones can be grouped into quadruples (relation, complement, inverse, inverse complement).

Group

A group is an algebraic structure consisting of a set together with an operation that combines any two of its elements to form a third element. To qualify as a group, the set and the operation must satisfy a few conditions called group axioms, namely closure, associativity, identity and invertibility.

Many familiar mathematical structures such as number systems obey these axioms: for example, the integers endowed with the addition operation form a group. However, the abstract formalisation of the group axioms, detached as it is from the concrete nature of any particular group and its operation, allows entities with highly diverse mathematical origins in abstract algebra and beyond to be handled in a flexible way, while retaining their essential structural aspects. The ubiquity of groups in numerous areas within and outside mathematics makes them a central organising principle of contemporary mathematics.

Groups share a fundamental kinship with the notion of symmetry. A symmetry group encodes symmetry features of a geometrical object: it consists of the set of transformations that leave the object unchanged, and the operation of combining two such transformations by performing one after the other. Such symmetry groups, particularly the continuous Lie groups, play an important role in many academic disciplines. Matrix groups, for example, can be used to understand fundamental physical laws underlying special relativity and symmetry phenomena in molecular chemistry.

The concept of a group arose from the study of polynomial equations, starting with Évariste Galois in the 1830s. After contributions from other fields such as number theory and geometry, the group notion was generalised and firmly established around 1870.

Modern group theory—a very active mathematical discipline—studies groups in their own right.[a[›]] To explore groups, mathematicians have devised various notions to break groups into smaller, better-understandable pieces, such as subgroups, quotient groups and simple groups.

In addition to their abstract properties, group theorists also study the different ways in which a group can be expressed concretely (its group representations), both from a theoretical and a computational point of view.

A particularly rich theory has been developed for finite groups, which culminated with the monumental classification of finite simple groups announced in 1983.[aa[›]] Since the mid-1980s, geometric group theory, which studies finitely generated groups as geometric objects, has become a particularly active area in group theory.

Definition and Illustration

First Example: the Integers: One of the most familiar groups is the set of integers Z which consists of the numbers

..., -4, -3, -2, -1, 0, 1, 2, 3, 4, ...

The following properties of integer addition serve as a model for the abstract group axioms given in the definition below.

1. For any two integers a and b, the sum $a + b$ is also an integer. Thus, adding two integers never yields some other type of number, such as a fraction. This property is known as *closure* under addition.
2. For all integers a, b and c, $(a + b) + c = a + (b + c)$. Expressed in words, adding a to b first, and then adding the result to c gives the same final result as adding a to the sum of b and c, a property known as *associativity*.
3. If a is any integer, then $0 + a = a + 0 = a$. Zero is called the *identity element* of addition because adding it to any integer returns the same integer.
4. For every integer a, there is an integer b such that $a + b = b + a = 0$. The integer b is called the *inverse element* of the integer a and is denoted $-a$.

The integers, together with the operation +, form a mathematical object belonging to a broad class sharing similar structural aspects. To appropriately understand these structures as a collective, the following abstract definition is developed.

Definition

A group is a set, G, together with an operation $\cdot$ (called the group law of G) that combines any two elements a and b to form another element, denoted $a \cdot b$ or ab. To qualify as a group, the set and operation, $(G, \cdot)$, must satisfy four requirements known as the group axioms:

Closure : For all a, b in G, the result of the operation, $a \cdot b$, is also in G.

Associativity : For all a, b and c in G, $(a \cdot b) \cdot c = a \cdot (b \cdot c)$.

Identity Element : There exists an element e in G, such that for every element a in G, the equation $e \cdot a = a \cdot e = a$ *holds*. The identity element of a group G is often written as 1 or 1_G, a notation inherited from the multiplicative identity.

Inverse Element: For each a in G, there exists an element b in G such that $a \cdot b = b \cdot a = 1_G$.

The order in which the group operation is carried out can be significant. In other words, the result of combining element a with element b need not yield the same result as combining element b with element a; the equation $a \cdot b = b \cdot a$ may not always be true. This equation does always hold in the group of integers under addition, because $a + b = b + a$ for any two integers (commutativity of addition). However, it does not always hold in the symmetry group below. Groups for which the equation $a \cdot b = b \cdot a$ always holds are called abelian (in honour of Niels Abel). Thus, the integer addition group is abelian, but the symmetry group in the following section is not.

The set G is called the *underlying set* of the group $(G, \cdot)$. Often the group's underlying set G is used as a short name for the group $(G, \cdot)$. Along the same lines, shorthand expressions such as "a subset of the group G″ or "an element of group G" are used when what is actually meant is "a subset of the underlying set G of the group $(G, \cdot)$″ or "an element of the underlying set G of the group $(G, \cdot)$″. Usually, it is clear from the context whether a symbol like G refers to a group or to an underlying set.

Second Example: a Symmetry Group

Two figures in the plane are congruent if one can be changed into the other using a combination of rotations, reflections, and translations. Any figure is congruent to itself. However, some figures are congruent to themselves in more than one way, and these extra congruences are called symmetries. A square has eight symmetries. These are:

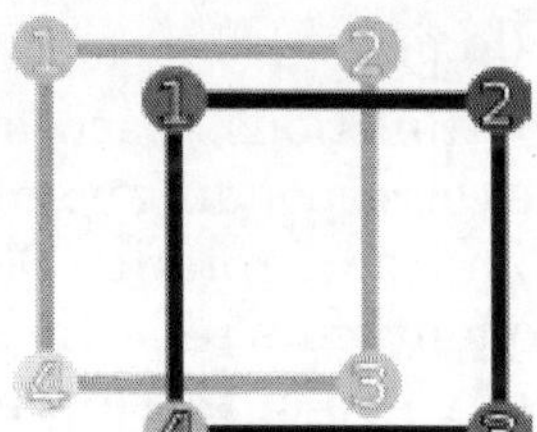

id (keeping it as is)

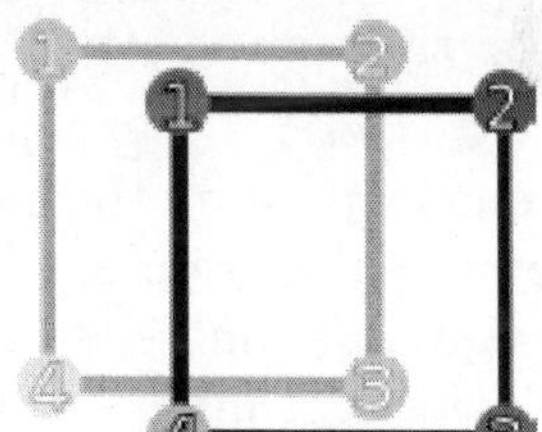

r_1 (rotation by 90° right)

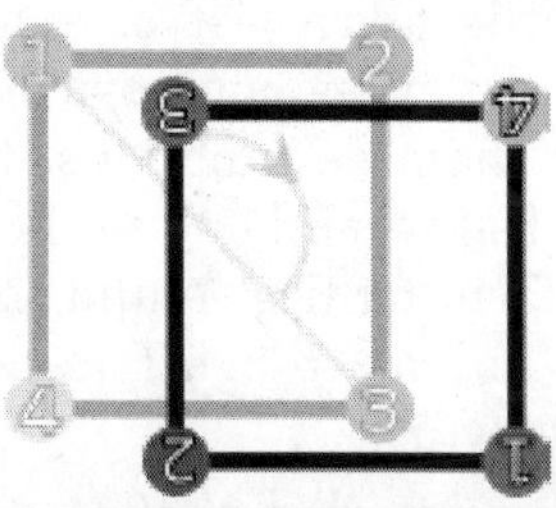

r_2 (rotation by 180° right)

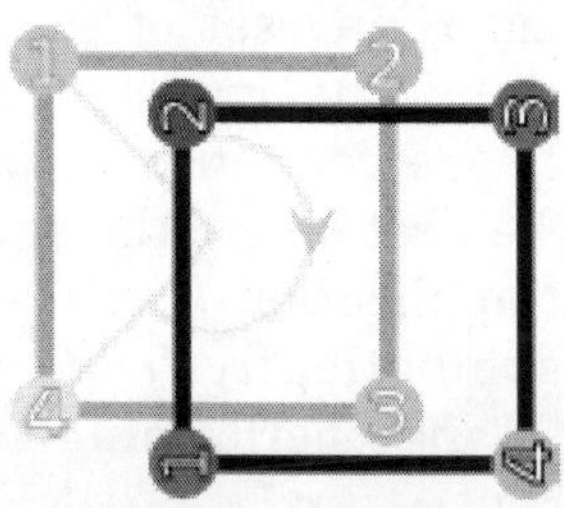

r_3 (rotation by 270° right)

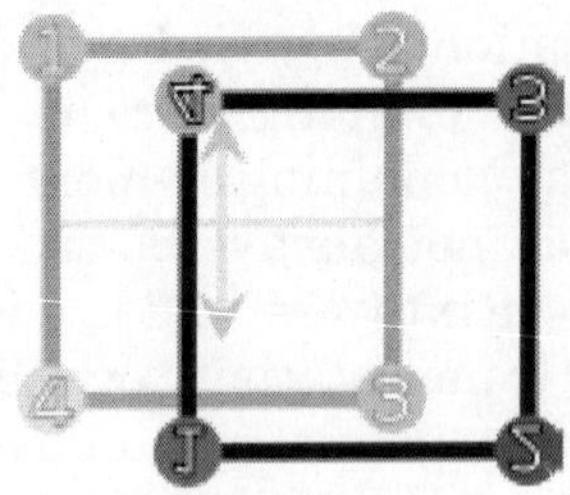

f_v (vertical flip)

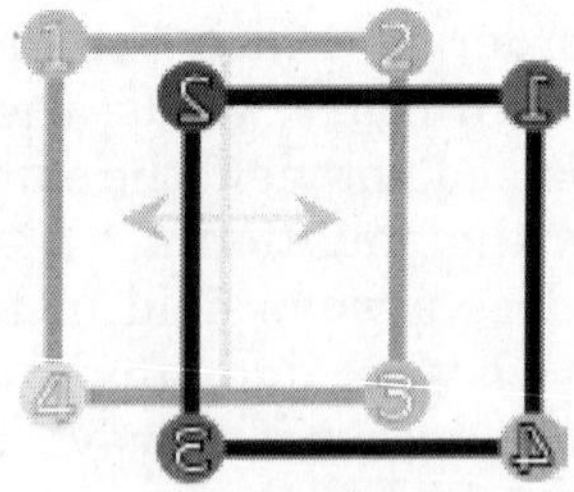

f_h (horizontal flip)

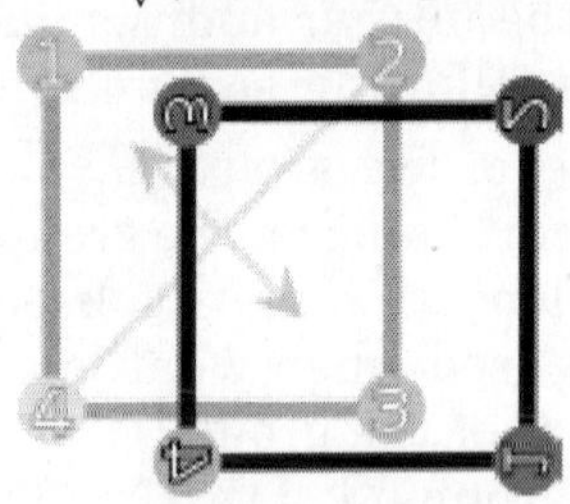

f_d (diagonal flip)

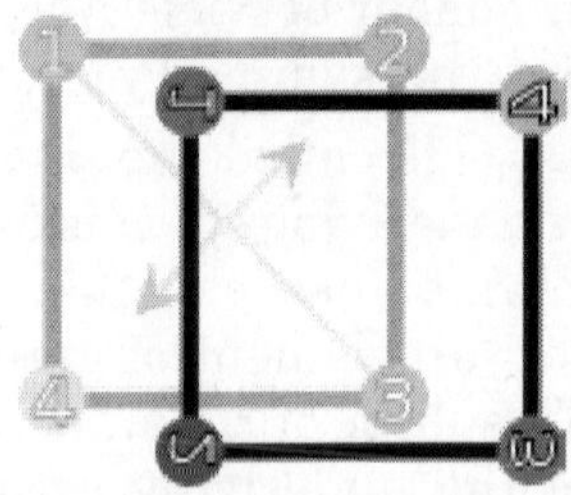

f_c (counter-diagonal flip)

Figure 1: The elements of the symmetry group of the square (D_4). The vertices are coloured and numbered only to visualise the operations.

- the identity operation leaving everything unchanged, denoted id;
- rotations of the square around its centre by 90° right, 180° right, and 270° right, denoted by r_1, r_2 and r_3, respectively;
- reflections about the vertical and horizontal middle line (f_h and f_v), or through the two diagonals (fd and f_c).

These symmetries are represented by functions. Each of these functions sends a point in the square to the corresponding point under the symmetry. For example, r_1 sends a point to its rotation 90° right around the square's centre, and f_h sends a point to its reflection across the square's vertical middle line. Composing two of these symmetry functions gives another symmetry function. These symmetries determine a group called the dihedral group of degree 4 and denoted D_4. The underlying set of the group is the above set of symmetry functions, and the group operation is function composition. Two symmetries are combined by composing them as functions, that is, applying the first one to the square, and the second one to the result of the first application. The result of performing first a and then b is written symbolically *from right to left* as $b \cdot a$ ("apply the symmetry b after performing the symmetry a").

The right-to-left notation is the same notation that is used for composition of functions.

The group table on the right lists the results of all such compositions possible. For example, rotating by 270° right (r_3) and then flipping horizontally (f_h) is the same as performing a reflection along the diagonal (f_d). Using the above symbols, highlighted in blue in the group table:

Group Table of D_4

•	Id	r_1	r_2	r_3	f_v	f_h	f_d	f_c
id	Id	r_1	r_2	r_3	f_v	f_h	f_d	f_c
r_1	r_1	r_2	r_3	id	f_c	f_d	f_v	f_h
r_2	r_2	r_3	id	r_1	f_h	f_v	f_c	f_d
r_3	r_3	id	r_1	r_2	f_d	f_c	f_h	f_v
f_v	f_v	f_d	f_h	f_c	id	r_2	r_1	r_3
f_h	f_h	f_c	f_v	f_d	r_2	id	r_3	r_1
f_d	f_d	f_h	f_c	f_v	r_3	r_1	id	r_2
f_c	f_c	f_v	f_d	f_h	r_1	r_3	r_2	id

The elements id, r_1, r_2, and r_3 form a subgroup, highlighted in red (upper left region). A left and right coset of this subgroup is highlighted in green (in the last row) and yellow (last column), respectively.

Given this set of symmetries and the described operation, the group axioms can be understood as follows:

1. The closure axiom demands that the composition $b \cdot a$ of any two symmetries $a$and b is also a symmetry. Another example for the group operation is

   $r_3 \cdot f_h = f_c$, i.e.

   rotating 270° right after flipping horizontally equals flipping along the counter-diagonal (f_c). Indeed every other combination of two symmetries still gives a symmetry, as can be checked using the group table.
2. The associativity constraint deals with composing more than two symmetries: Starting with three elements a, b and c of D_4, there are two possible ways of using these three symmetries in this order to determine a symmetry of the square. One of these ways is to first compose a and b into a single symmetry, then to compose that symmetry with c. The other way is to first compose b and c, then to compose the resulting symmetry with a. The associativity condition

   $(a \cdot b) \cdot c = a \cdot (b \cdot c)$

means that these two ways are the same, i.e., a product of many group elements can be simplified in any order. For example, $(f_d \cdot f_v) \cdot r_2 = f_d \cdot (f_v \cdot r_2)$ can be checked using the group table in the below

$(f_d \cdot f_v) \cdot r_2$	=	$r_3 \cdot r_2$	=	r_1, which equals
$f_d \cdot (f_v \cdot r_2)$	=	$f_d \cdot f_h$	=	r_1.

While associativity is true for the symmetries of the square and addition of numbers, it is not true for all operations. For instance, subtraction of numbers is not associative: (7 -3) -2 = 2 is not the same as 7 - (3 - 2) = 6.

1. The identity element is the symmetry id leaving everything unchanged: for any symmetry a, performing id after a (or a after id) equals a, in symbolic form,

   $\text{id} \cdot a = a,$

   $a \cdot \text{id} = a.$

2. An inverse element undoes the transformation of some other element. Every symmetry can be undone: each of the following transformations—identity id, the flips f_h, f_v, f_d, f_c and the 180° rotation r_2—is its own inverse, because performing it twice brings the square back to its original orientation. The rotations r_3 and r_1 are each other's inverses, because rotating 90° and then rotation 270° (or vice versa) yields a rotation over 360° which leaves the square unchanged. In symbols,

   $f_h \cdot f_h = \text{id},$

   $r_3 \cdot r_1 = r_1 \cdot r_3 = \text{id}.$

In contrast to the group of integers above, where the order of the operation is irrelevant, it does matter in D_4: $f_h \cdot r_1 = f_c$ but $r_1 \cdot f_h = f_d$. In other words, D_4 is not abelian, which makes the group structure more difficult than the integers introduced first.

History

The modern concept of an abstract group developed out of several fields of mathematics. The original motivation for group theory was the quest for solutions of polynomial equations of degree higher than 4. The 19th-century French mathematician Évariste Galois, extending prior work of Paolo Ruffini and Joseph-Louis Lagrange, gave a criterion for the solvability of a particular polynomial equation in terms of the symmetry group of its roots (solutions). The elements of such a Galois group correspond to certain permutations of the roots. At first, Galois' ideas were rejected by his contemporaries, and published only posthumously. More general permutation groups were investigated in particular by Augustin Louis Cauchy. Arthur Cayley's On the theory of groups, as depending on the symbolic equation $\theta^n = 1$ (1854) gives the first abstract definition of a finite group.

Geometry was a second field in which groups were used systematically, especially symmetry groups as part of Felix Klein's 1872 Erlangen program. After novel geometries such as hyperbolic and projective geometry had emerged, Klein used group theory to organise them in a more coherent way. Further advancing these ideas, Sophus Lie founded the study of Lie groups in 1884. The third field contributing to group theory was number theory. Certain abelian group structures had been used implicitly in Carl Friedrich Gauss' number-theoretical work Disquisitiones Arithmeticae(1798), and more explicitly by Leopold Kronecker. In 1847, Ernst Kummer led early attempts to prove Fermat's Last Theorem to a climax by developing groups describing factorisation into prime numbers.

The convergence of these various sources into a uniform theory of groups started with Camille Jordan's Traité des substitutions et des équations algébriques (1870). Walther von Dyck (1882) gave the first statement of the modern definition of an abstract group. As of the 20th century, groups gained wide recognition by the pioneering work of Ferdinand Georg Frobenius and William Burnside, who worked on representation theory of finite groups, Richard Brauer's modular representation theory and Issai Schur's papers. The theory of Lie groups, and more generally locally compact groups was pushed by Hermann Weyl, Élie Cartan and many others. Its algebraic counterpart, the theory of algebraic groups, was first shaped by Claude Chevalley (from the late 1930s) and later by pivotal work of Armand Borel and Jacques Tits.

The University of Chicago's 1960–61 Group Theory Year brought together group theorists such as Daniel Gorenstein, John G. Thompson and Walter Feit, laying the foundation of a collaboration that, with input from numerous other mathematicians, classified all finite simple groups in 1982. This project exceeded previous mathematical endeavours by its sheer size, in both length of proof and number of researchers. Research is ongoing to simplify the proof of this classification. These days, group theory is still a highly active mathematical branch crucially impacting many other fields.[a[b]]

Elementary Consequences of the Group Axioms

Basic facts about all groups that can be obtained directly from the group axioms are commonly subsumed under elementary group theory. *For example, repeated applications of the associativity axiom show that the unambiguity of*

$$a \cdot b \cdot c = (a \cdot b) \cdot c = a \cdot (b \cdot c)$$

generalises to more than three factors. Because this implies that parentheses can be inserted anywhere within such a series of terms, parentheses are usually omitted.

The axioms may be weakened to assert only the existence of a left identity and left inverses. Both can be shown to be actually two-sided, so the resulting definition is equivalent to the one given above.

Uniqueness of Identity Element and Inverses

Two important consequences of the group axioms are the uniqueness of the identity element and the uniqueness of inverse elements. There can be only one identity element in a group, and each element in a group has exactly one inverse element. Thus, it is customary to speak of *the* identity, and *the inverse* of an element.

To prove the uniqueness of an inverse element of a, suppose that a has two inverses, denoted l and r, in a group $(G, \cdot)$. Then

$l = l \cdot 1_G$	as 1_G is the identity element
$= l \cdot (a \cdot r)$	because r is an inverse of a, so $1_G = a \cdot r$
$= (l \cdot a) \cdot r$	by associativity, which allows to rearrange the parentheses
$= 1_G \cdot r$	Since l is an inverse of a, i.e. $l \cdot a = 1_G$
$= r$	for 1_G is the identity element

The two extremal terms l and r are equal, since they are connected by a chain of equalities. In other words there is only one inverse element of a. Similarly, to prove that the identity element of a group is unique, assume G is a group with two identity elements 1_G and e. Then $1_G = 1_G \cdot e = e$, hence *1G and e* are equal.

Division

In groups, it is possible to perform division: given elements a and b of the group G, there is exactly one solution x in G to the equation $x \cdot a = b$. In fact, right multiplication of the equation by a^{-1} gives the solution $x = x \cdot a \cdot a^{-1} = b \cdot a^{-1}$. Similarly there is exactly one solution y in G to the equation$a \cdot y = b$, namely $y = a^{-1} \cdot b$. In general, x and y need not agree.

A consequence of this is that multiplying by a group element g is a bijection. Specifically, if g is an element of the group G, there is a bijection from G to itself called left translation by g sending h - $G$to $g \cdot h$. Similarly, right translation by g is a bijection from G to itself sending h to $h \cdot g$. If G is abelian, left and right translation by a group element are the same.

Basic Concepts

The following sections use mathematical symbols such as $X = \{x, y, z\}$ *to denote a set X containing elements x, y, and z, or alternatively x " X to restate that x is an element of X. The notation* $f: X \to Y$ *means f is a function assigning to every element of X an element of Y.*

To understand groups beyond the level of mere symbolic manipulations as above, more structural concepts have to be employed.[c[›]] There is a conceptual principle underlying all of the following notions: to take advantage of the structure offered by groups (which sets, being "structureless", do not have), constructions related to groups have to be *compatible* with the group operation. This compatibility manifests itself in the following notions in various ways. For example, groups can be related to each other via functions called group homomorphisms. By the mentioned principle, they are required to respect the group structures in a precise sense. The structure of groups can also be understood by breaking them into pieces called subgroups and quotient groups. The principle of "preserving structures"—a recurring topic in mathematics throughout—is an instance of working in a category, in this case the category of groups.

Group Homomorphisms

Group homomorphisms[g[›]] are functions that preserve group structure. A function $a: G \to H$ between two groups $(G,\cdot)$ and $(H,*)$ is called a *homomorphism* if the equation

$$a(g \cdot k) = a(g) * a(k)$$

holds for all elements g, k in G. In other words, the result is the same when performing the group operation after or before applying the map a. This requirement ensures that $a(1_G) = 1_H$, and also$a(g)^{-1} = a(g^{-1})$ for all g in G. Thus a group homomorphism respects all the structure of G provided by the group axioms.

Two groups G and H are called *isomorphic* if there exist group homomorphisms $a: G \to H$ and b: $H \to G$, such that applying the two functions one after another in each of the two possible orders gives the identity functions of G and H. That is, $a(b(h)) = h$ and $b(a(g)) = g$ for any g in G and h in H. From an abstract point of view, isomorphic groups carry the same information. For example, proving that $g \cdot g = 1_G$ for some element g of G is equivalent to proving that $a(g) * a(g) = 1_H$, because applying a to the first equality yields the second, and applying b to the second gives back the first.

Subgroups

Informally, a *subgroup* is a group H contained within a bigger one, G. Concretely, the identity element of G is contained in H, and whenever h_1 and h_2 are in H, then so are $h_1 \cdot h_2$ and h_1^{-1}, so the elements of H, equipped with the group operation on G restricted to H, indeed form a group.

In the example above, the identity and the rotations constitute a subgroup R = {id, r_1, r_2, r_3}, highlighted in red in the group table above: any two rotations composed are still a rotation, and a rotation can be undone by (i.e. is inverse to) the complementary rotations 270° for 90°, 180° for 180°, and 90° for 270° (note that rotation in the opposite direction is not defined). The subgroup test is a necessary and sufficient condition for a subset H of a group G to be a subgroup: it is sufficient to check that $g^{-1}h$ - H for all elements g, h - H. Knowing the subgroups is important in understanding the group as a whole.[d[?]]

Given any subset S of a group G, the subgroup generated by S consists of products of elements of $S$and their inverses. It is the smallest subgroup of G containing S. In the introductory example above, the subgroup generated by r_2 and f_v consists of these two elements, the identity element id and $f_h = f_v \cdot r_2$. Again, this is a subgroup, because combining any two of these four elements or their inverses (which are, in this particular case, these same elements) yields an element of this subgroup.

Cosets

In many situations it is desirable to consider two group elements the same if they differ by an element of a given subgroup. For example, in D_4 above, once a flip is performed, the square never gets back to the r_2 configuration by just applying the rotation operations (and no further flips), i.e. the rotation operations are irrelevant to the question whether a flip has been performed. Cosets are used to formalise this insight: a subgroup H defines left and right cosets, which can be thought of as translations of H by arbitrary group elements g. In symbolic terms, the *left* and *right* cosets of H containing g are

$gH = \{g \cdot h, h \in H\}$ and $Hg = \{h \cdot g, h \in H\}$, respectively.

The cosets of any subgroup H form a partition of G; that is, the union of all left cosets is equal to G and two left cosets are either equal or have an empty intersection. The first case $g_1H = g_2H$ happens precisely when $g_1^{-1} \cdot g_2$ - H, i.e. if the two elements differ by an element of H. Similar considerations apply to the right cosets of H. The left and right cosets of H may or may not be equal. If they are, i.e. for all g in G, $gH = Hg$, then H is said to be a *normal subgroup*.

In D_4, the introductory symmetry group, the left cosets gR of the subgroup R consisting of the rotations are either equal to R, if g is an element of R itself, or otherwise equal to $U = f_cR = \{f_c, f_v, f_d, f_h\}$ (highlighted in green). The subgroup R is also normal, because $f_cR = U = Rf_c$ and similarly for any element other than f_c.

Quotient Groups

In some situations the set of cosets of a subgroup can be endowed with a group law, giving a *quotient group* or *factor group*. For this to be possible, the subgroup has to be normal. Given any normal subgroup N, the quotient group is defined by

$G / N = \{gN, g \in G\}$, □ G modulo N".

This set inherits a group operation (sometimes called coset multiplication, or coset addition) from the original group G: $(gN) \cdot (hN) = (gh)N$ for all g and h in G. This definition is motivated by the idea (itself an instance of general structural considerations outlined above) that the map $G \to G / N$ that associates to any element g its coset gN be a group homomorphism, or by general abstract considerations called universal properties. The coset $eN = N$ serves as the identity in this group, and the inverse of gN in the quotient group is $(gN)^{-1} = (g^{-1})N$.[e[b]]

•	R	U
R	R	U
U	U	R

Group table of the quotient group D_4 / R.

The elements of the quotient group D_4 / R are R itself, which represents the identity, and $U = f_vR$. The group operation on the quotient is shown at the right. For example, $U \cdot U = f_vR \cdot f_vR = (f_v \cdot f_v)R = R$. Both the subgroup $R = \{id, r_1, r_2, r_3\}$, as well as the corresponding quotient are abelian, whereas D_4 is not abelian. Building bigger groups by smaller ones, such as D4 from its subgroup R and the quotient D_4 / R is abstracted by a notion called semidirect product.

Quotient groups and subgroups together form a way of describing every group by its *presentation*: any group is the quotient of the free group over the *generators* of the group, quotiented by the subgroup of relations. The dihedral group D_4, for example, can be generated by two elements r and f (for example, $r = r_1$, the right rotation and $f = f_v$ the vertical (or any other) flip), which means that every symmetry of the square is a finite composition of these two symmetries or their inverses. Together with the relations

$$r^4 = f^2 = (r \cdot f)^2 = 1,$$

the group is completely described. A presentation of a group can also be used to construct the Cayley graph, a device used to graphically capture discrete groups.

Sub- and quotient groups are related in the following way: a subset H of G can be seen as an injectivemap $H \to G$, i.e. any element of the target has at most one element that maps to it. The counterpart to injective maps are surjective maps (every element of the target is mapped onto), such as the canonical map $G \to G / N$.[y[›]] Interpreting subgroup and quotients in light of these homomorphisms emphasizes the structural concept inherent to these definitions alluded to in the introduction. In general, homomorphisms are neither injective nor surjective. Kernel and image of group homomorphisms and the first isomorphism theorem address this phenomenon.

Examples and Applications

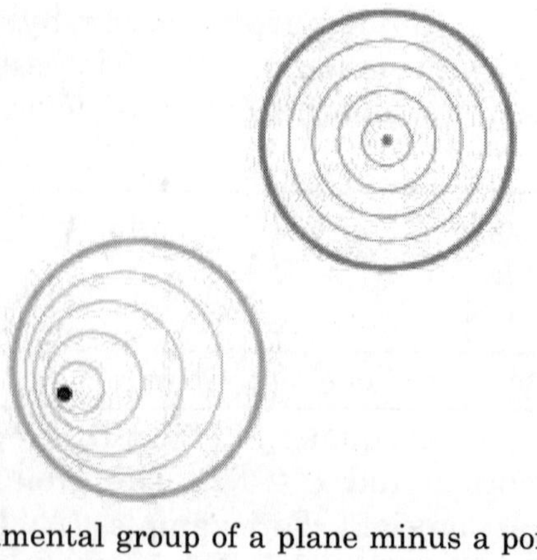

Figure 2: The fundamental group of a plane minus a point (bold) consists of loops around the missing point. This group is isomorphic to the integers.

Examples and applications of groups abound. A starting point is the group Z of integers with addition as group operation, introduced above. If instead of addition multiplication is considered, one obtains multiplicative groups. These groups are predecessors of important constructions in abstract algebra.

Groups are also applied in many other mathematical areas. Mathematical objects are often examined by associating groups to them and studying the properties of the corresponding groups. For example, Henri Poincaré founded what is now called algebraic topology by introducing the fundamental group. By means of this connection, topological properties such as proximity and continuity translate into properties of groups.[i[›]] For example, elements of the fundamental group are represented by loops. The second image at the right shows some loops in a plane minus a point. The blue loop is considered null-homotopic (and thus irrelevant), because it can be continuously shrunk to a point. The presence of the hole prevents the orange loop from being shrunk to a point. The fundamental group of the plane with a point deleted turns out to be infinite cyclic, generated by the orange loop (or any other loop winding once around the hole). This way, the fundamental group detects the hole.

In more recent applications, the influence has also been reversed to motivate geometric constructions by a group-theoretical background.[j[›]] In a similar vein, geometric group theory employs geometric concepts, for example in the study of hyperbolic groups. Further branches crucially applying groups include algebraic geometry and number theory.

In addition to the above theoretical applications, many practical applications of groups exist. Cryptography relies on the combination of the abstract group theory approach together with algorithmical knowledge obtained in computational group theory, in particular when implemented for finite groups. Applications of group theory are not restricted to mathematics; sciences such as physics, chemistry and computer science benefit from the concept.

Numbers

Many number systems, such as the integers and the rationals enjoy a naturally given group structure. In some cases, such as with the rationals, both addition and multiplication operations give rise to group structures. Such number systems are predecessors to more general algebraic structures known as rings and fields. Further abstract algebraic concepts such as modules, vector spaces and algebras also form groups.

Integers

The group of integers Z under addition, denoted (Z, +), has been described above. The integers, with the operation of multiplication instead of addition, (Z, ·) do *not* form a group. The closure, associativity and identity axioms are satisfied, but inverses do not exist: for example, $a = 2$ is an integer, but the only solution to the equation $a \cdot b = 1$ in this case is $b = 1/2$, which is a rational number, but not an integer. Hence not every element of Z has a (multiplicative) inverse.[k]

Rationals

The desire for the existence of multiplicative inverses suggests considering fractions

$$\frac{a}{b}$$

Fractions of integers (with b nonzero) are known as rational numbers.[l] The set of all such fractions is commonly denoted Q. There is still a minor obstacle for (Q, ·), the rationals with multiplication, being a group: because the rational number 0 does not have a multiplicative inverse (i.e., there is no x such that $x \cdot 0 = 1$), (Q, ·) is still not a group.

However, the set of all *nonzero* rational numbers $Q \setminus \{0\} = \{q \in Q, q \neq 0\}$ does form an abelian group under multiplication, denoted $(Q \setminus \{0\}, \cdot)$.[m] Associativity and identity element axioms follow from the properties of integers. The closure requirement still holds true after removing zero, because the product of two nonzero rationals is never zero. Finally, the inverse of a/b is b/a, therefore the axiom of the inverse element is satisfied.

The rational numbers (including 0) also form a group under addition. Intertwining addition and multiplication operations yields more complicated structures called rings and—if division is possible, such as in Q—fields, which occupy a central position in abstract algebra. Group theoretic arguments therefore underlie parts of the theory of those entities.[n]

Nonzero Integers Modulo a Prime

For any prime number p, modular arithmetic furnishes the multiplicative group of integers modulo p. Its elements are integers not divisible by p, considered modulo p, i.e. two numbers are considered equivalent if their difference is divisible by p. For example, if $p = 5$, there are exactly four group elements 1, 2, 3, 4: multiples of 5 are excluded and 6 and “4 are both equivalent to 1 etc. The group operation is given by multiplication. Therefore, $4 \cdot 4 = 1$, because the usual product 16 is equivalent to 1, for 5 divides $16 - 1 = 15$, denoted

$$16 \equiv 1 \pmod 5.$$

The primality of p ensures that the product of two integers neither of which is divisible by p is not divisible by p either, hence the indicated set of classes is closed under multiplication.[o] The identity element is 1, as usual for a multiplicative group, and the associativity follows from the corresponding property of integers. Finally, the inverse element axiom requires that given an integer a not divisible by p, there exists an integer b such that

$$a \cdot b \equiv 1 \pmod p, \text{ i.e. } p \text{ divides the difference } a \cdot b - 1.$$

The inverse b can be found by using Bézout's identity and the fact that the greatest common divisor gcd(a, p) equals 1. In the case p = 5 above, the inverse of 4 is 4, and the inverse of 3 is 2, as $3 \cdot 2 = 6 \equiv 1 \pmod 5$. Hence all group axioms are fulfilled. Actually, this example is similar to $(Q\backslash\{0\}, \cdot)$ above, because it turns out to be the multiplicative group of nonzero elements in the finite field F_p, denoted $F_p^{\times}$. These groups are crucial to public-key cryptography.[p]

Cyclic Groups

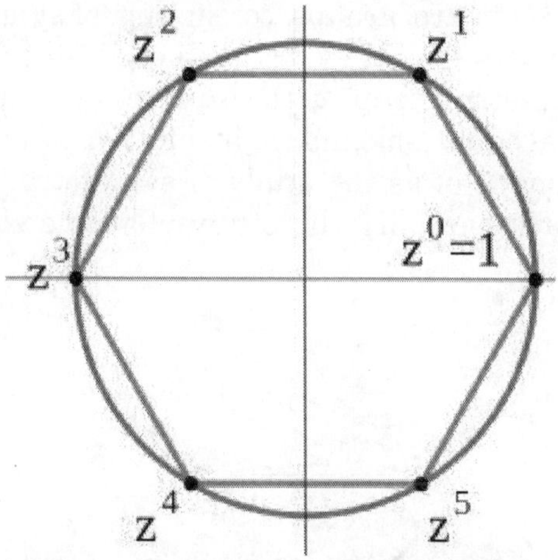

Figure 3: The 6th complex roots of unity form a cyclic group. z is a primitive element, but z^2 is not, because the odd powers of z are not a power of z^2.

A *cyclic group* is a group all of whose elements are powers (when the group operation is written additively, the term 'multiple' can be used) of a particular element a. In multiplicative notation, the elements of the group are:

$$\ldots, a^{-3}, a^{-2}, a^{-1}, a^0 = e, a, a^2, a^3, \ldots,$$

where a^2 means $a \cdot a$, and a^{-3} stands for $a^{-1} \cdot a^{-1} \cdot a^{-1} = (a \cdot a \cdot a)^{-1}$ etc.[h[?]] Such an element a is called a generator or a primitive element of the group.

A typical example for this class of groups is the group of n-th complex roots of unity, given by complex numbers z satisfying $z^n = 1$ (and whose operation is multiplication). Any cyclic group with n elements is isomorphic to this group. Using some field theory, the group $F_p^\times$ can be shown to be cyclic: for example, if $p = 5$, 3 is a generator since $3^1 = 3$, $3^2 = 9 \equiv 4$, $3^3 \equiv 2$, and $3^4 \equiv 1$.

Some cyclic groups have an infinite number of elements. In these groups, for every non-zero element a, all the powers of a are distinct; despite the name "cyclic group", the powers of the elements do not cycle. An infinite cyclic group is isomorphic to (Z, +), the group of integers under addition introduced above. As these two prototypes are both abelian, so is any cyclic group.

The study of abelian groups is quite mature, including the fundamental theorem of finitely generated abelian groups; and reflecting this state of affairs, many group-related notions, such as centre and commutator, describe the extent to which a given group is not abelian.

Symmetry Groups

Symmetry groups are groups consisting of symmetries of given mathematical objects—be they of geometric nature, such as the introductory symmetry group of the square, or of algebraic nature, such as polynomial equations and their solutions. Conceptually, group theory can be thought of as the study of symmetry.[t[?]] Symmetries in mathematics greatly simplify the study of geometrical or analytical objects.

Figure 4: Rotations and flips form the symmetry group of a great icosahedron.

A group is said to act on another mathematical object X if every group element performs some operation on X compatibly to the group law. In the rightmost example below, an element of order 7 of the (2,3,7) triangle group acts on the tiling by permuting the highlighted warped triangles (and the other ones, too). By a group action, the group pattern is connected to the structure of the object being acted on. In chemical fields, such as crystallography, space groups andpoint groups describe molecular symmetries and crystal symmetries. These symmetries underlie the chemical and physical behavior of these systems, and group theory enables simplification of quantum mechanical analysis of these properties. For example, group theory is used to show that optical transitions between certain quantum levels cannot occur simply because of the symmetry of the states involved.

Not only are groups useful to assess the implications of symmetries in molecules, but surprisingly they also predict that molecules sometimes can change symmetry. The Jahn-Teller effect is a distortion of a molecule of high symmetry when it adopts a particular ground state of lower symmetry from a set of possible ground states that are related to each other by the symmetry operations of the molecule.

Likewise, group theory helps predict the changes in physical properties that occur when a material undergoes a phase transition, for example, from a cubic to a tetrahedral crystalline form. An example is ferroelectric materials, where the change from a paraelectric to a ferroelectric state occurs at theCurie temperature and is related to a change from the high-symmetry paraelectric state to the lower symmetry ferroelectic state, accompanied by a so-called soft phonon mode, a vibrational lattice mode that goes to zero frequency at the transition. Such spontaneous symmetry breaking has found further application in elementary particle physics, where its occurrence is related to the appearance of Goldstone bosons.

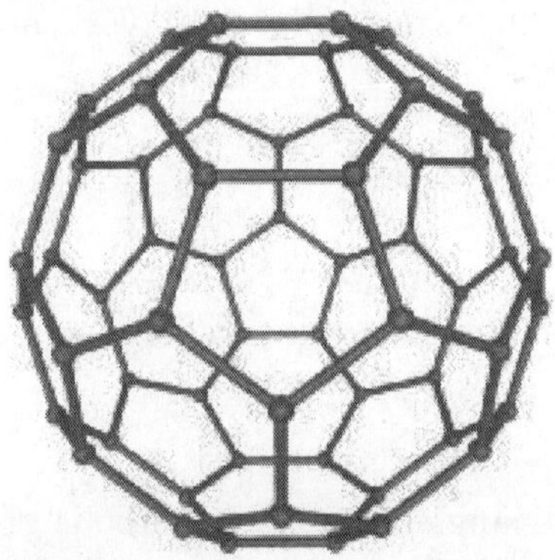

Figure 5: Buckminster fullerene displays icosahedral symmetry.

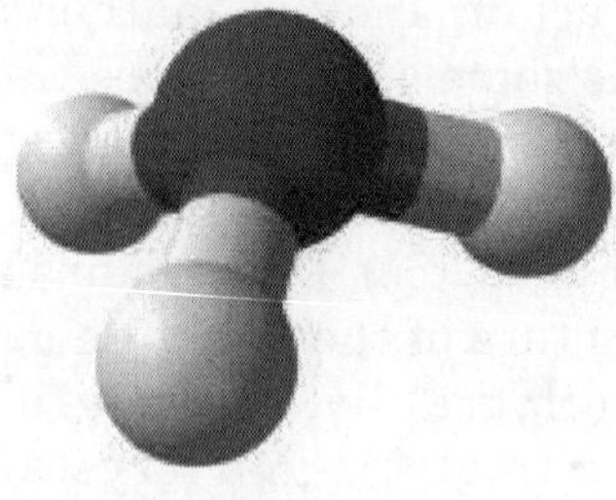

Figure 6: Ammonia, NH_3. Its symmetry group is of order 6, generated by a 120° rotation and a reflection.

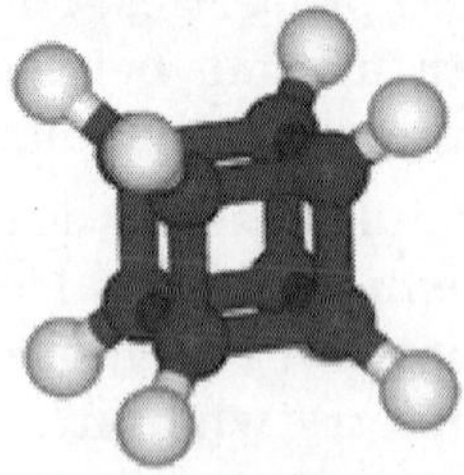

Figure 7: Cubane C_8H_8features octahedral symmetry.

Figure 8: Hexaaquacopper(II) complexion, [Cu(OH2)6]2+. Compared to a perfectly symmetrical shape, the molecule is vertically dilated by about 22% (Jahn-Teller effect).

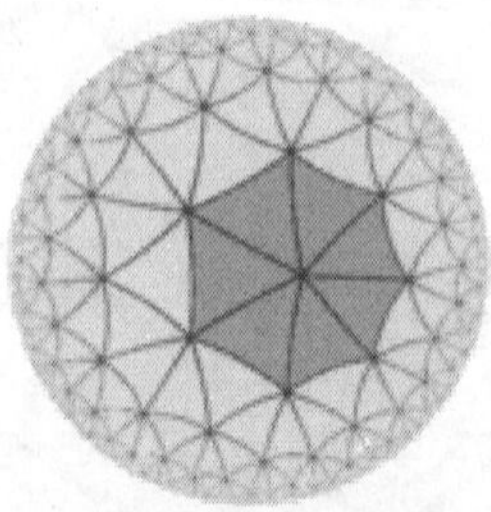

Figure 9: The (2,3,7) triangle group, a hyperbolic group, acts on this tiling of the hyperbolic plane.

Finite symmetry groups such as the Mathieu groups are used in coding theory, which is in turn applied in error correction of transmitted data, and in CD players.

Another application is differential Galois theory, which characterises functions having antiderivatives of a prescribed form, giving group-theoretic criteria for when solutions of certain differential equations are well-behaved.[u[?]] Geometric properties that remain stable under group actions are investigated in (geometric) invariant theory.

General Linear Group and Representation Theory

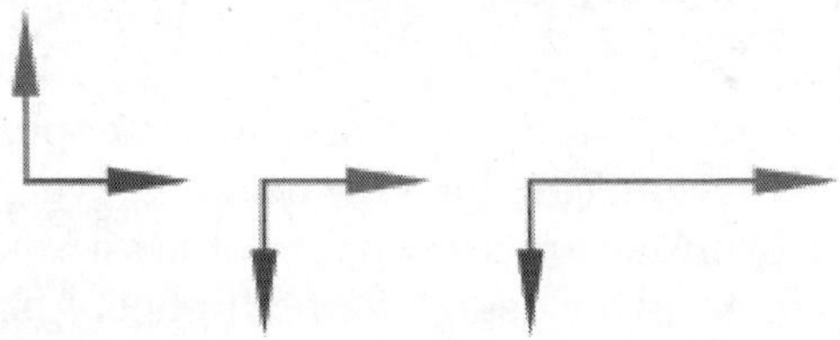

Figure 10: Two vectors (the left illustration) multiplied by matrices (the middle and right illustrations). The middle illustration represents a clockwise rotation by 90°, while the right-most one stretches the x-coordinate by factor 2.

Matrix groups consist of matrices together with matrix multiplication. The *general linear group* $GL(n, \mathrm{R})$consists of all invertible n-by-n matrices with real entries. Its subgroups are referred to as *matrix groups* or *linear groups*. The dihedral group example mentioned above can be viewed as a (very small) matrix group. Another important matrix group is the special orthogonal group $SO(n)$. It describes all possible rotations in n dimensions. Via Euler angles, rotation matrices are used in computer graphics.

Representation theory is both an application of the group concept and important for a deeper understanding of groups. It studies the group by its group actions on other spaces. A broad class of group representations are linear representations, i.e. the group is acting on a vector space, such as the three-dimensional Euclidean space R^3. A representation of G on an n-dimensional real vector space is simply a group homomorphism

$$\rho: G \to GL(n, \mathrm{R})$$

from the group to the general linear group. This way, the group operation, which may be abstractly given, translates to the multiplication of matrices making it accessible to explicit computations.[w[?]]

Given a group action, this gives further means to study the object being acted on.[x[b]] On the other hand, it also yields information about the group. Group representations are an organising principle in the theory of finite groups, Lie groups, algebraic groups and topological groups, especially (locally) compact groups.

Galois Groups

Galois groups have been developed to help solve polynomial equations by capturing their symmetry features. For example, the solutions of the quadratic equation $ax^2 + bx + c = 0$ are given by

$$x = \frac{-b \pm \sqrt{b^2 - 4ac}}{2a}.$$

Exchanging "+" and "–" in the expression, i.e. permuting the two solutions of the equation can be viewed as a (very simple) group operation. Similar formulae are known for cubic and quartic equations, but do *not* exist in general for degree 5 and higher. Abstract properties of Galois groups associated with polynomials (in particular their solvability) give a criterion for polynomials that have all their solutions expressible by radicals, i.e. solutions expressible using solely addition, multiplication, and roots similar to the formula above.

The problem can be dealt with by shifting to field theory and considering the splitting field of a polynomial. Modern Galois theory generalises the above type of Galois groups to field extensions and establishes—via the fundamental theorem of Galois theory—a precise relationship between fields and groups, underlining once again the ubiquity of groups in mathematics.

Finite Groups

A group is called *finite* if it has a finite number of elements. The number of elements is called the order of the group. An important class is the *symmetric groups* S_N, the groups of permutations of *N*l etters. For example, the symmetric group on 3 letters S_3 is the group consisting of all possible orderings of the three letters *ABC*, i.e. contains the elements *ABC*, *ACB*, ..., up to *CBA*, in total 6 (or 3 factorial) elements. This class is fundamental insofar as any finite group can be expressed as a subgroup of a symmetric group S_N for a suitable integer *N* (Cayley's theorem). Parallel to the group of symmetries of the square above, S_3 can also be interpreted as the group of symmetries of an equilateral triangle.

The order of an element a in a group G is the least positive integer n such that $a^n = e$, *where* a^n *represents*

$$\underbrace{a \cdots a}_{n \text{ factors}},$$

i.e. application of the operation • to n copies of a. (If • represents multiplication, then a^n corresponds to the nth power of a.) In infinite groups, such an n may not exist, in which case the order of a is said to be infinity. The order of an element equals the order of the cyclic subgroup generated by this element.

More sophisticated counting techniques, for example counting cosets, yield more precise statements about finite groups: Lagrange's Theorem states that for a finite group *G* the order of any finite subgroup *H* divides the order of *G*. The Sylow theorems give a partial converse.

The dihedral group is a finite group of order 8. The order of r_1 is 4, as is the order of the subgroup *R* it generates. The order of the reflection elements f_v etc. is 2. Both orders divide 8, as predicted by Lagrange's Theorem. The groups $F_p^{\times}$ above have order p - 1.

Classification of Finite Simple Groups

Mathematicians often strive for a complete classification (or list) of a mathematical notion. In the context of finite groups, this aim quickly leads to difficult and profound mathematics. According to Lagrange's theorem, finite groups of order *p*, a prime number, are necessarily cyclic (abelian) groups Z_p. Groups of order p^2 can also be shown to be abelian, a statement which does not generalise to order p^3, as the non-abelian group D_4 of order 8 = 2^3 above shows. Computer algebra systems can be used to list small groups, but there is no classification of all finite groups.[q[›]] An intermediate step is the classification of finite simple groups.[r[›]] A nontrivial group is called *simple* if its only normal subgroups are the trivial group and the group itself.[s[›]] The Jordan–Hφlder theorem exhibits finite simple groups as the building blocks for all finite groups. Listing all finite simple groups was a major achievement in contemporary group theory. 1998 Fields Medal winner Richard Borcherds succeeded to prove the monstrous moonshine conjectures, a surprising and deep relation of the largest finite simple sporadic group—the "monster group"—with certain modular functions, a piece of classical complex analysis, and string theory, a theory supposed to unify the description of many physical phenomena.

Groups with Additional Structure

Many groups are simultaneously groups and examples of other mathematical structures. In the language of category theory, they are group objects in a category, meaning that they are objects (that is, examples of another mathematical structure) which come with transformations (called morphisms) that mimic the group axioms. For example, every group (as defined above) is also a set, so a group is a group object in the category of sets.

Topological Groups

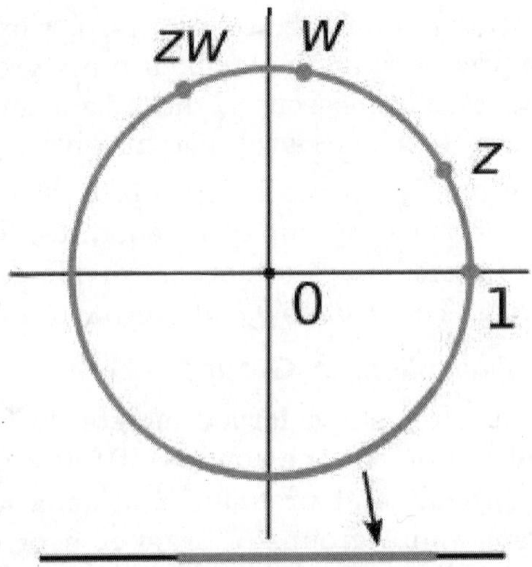

Figure 11: The unit circle in the complex plane under complex multiplication is a Lie group and, therefore, a topological group. It is topological since complex multiplication and division are continuous. It is a manifold and thus a Lie group, because every small piece, such as the red arc in the figure, looks like a part of the real line (shown at the bottom).

Some topological spaces may be endowed with a group law. In order for the group law and the topology to interweave well, the group operations must be continuous functions, that is, $g \cdot h$, and g^{-1} must not vary wildly if *g*a nd *h* vary only little.

Such groups are called *topological groups,* and they are the group objects in the category of topological spaces. The most basic examples are the reals R under addition, (R \ {0}, ·), and similarly with any other topological field such as the complex numbers or *p*-adic numbers. All of these groups are locally compact, so they have Haar measures and can be studied via harmonic analysis. The former offer an abstract formalism of invariant integrals. Invariance means, in the case of real numbers for example:

$$\int f(x)dx = \int f(x+c)dx$$

for any constant c. Matrix groups over these fields fall under this regime, as do adele rings and adelic algebraic groups, which are basic to number theory.

Galois groups of infinite field extensions such as the absolute Galois group can also be equipped with a topology, the so-called Krull topology, which in turn is central to generalise the above sketched connection of fields and groups to infinite field extensions. An advanced generalisation of this idea, adapted to the needs of algebraic geometry, is the étale fundamental group.

Lie Groups

Lie groups (in honour of Sophus Lie) are groups which also have a manifold structure, i.e. they are spaces looking locally like some Euclidean space of the appropriate dimension. Again, the additional structure, here the manifold structure, has to be compatible, i.e. the maps corresponding to multiplication and the inverse have to be smooth.

A standard example is the general linear group introduced above: it is an open subset of the space of all n-by-n matrices, because it is given by the inequality

$$\det(A) \neq 0,$$

where A denotes an n-by-n matrix.

Lie groups are of fundamental importance in physics: Noether's theorem links continuous symmetries to conserved quantities. Rotation, as well as translations in space and time are basic symmetries of the laws of mechanics. They can, for instance, be used to construct simple models—imposing, say, axial symmetry on a situation will typically lead to significant simplification in the equations one needs to solve to provide a physical description.[v[›]] Another example are the Lorentz transformations, which relate measurements of time and velocity of two observers in motion relative to each other.

They can be deduced in a purely group-theoretical way, by expressing the transformations as a rotational symmetry of Minkowski space. The latter serves—in the absence of significant gravitation—as a model of space time in special relativity. The full symmetry group of Minkowski space, i.e. including translations, is known as the Poincaré group. By the above, it plays a pivotal role in special relativity and, by implication, for quantum field theories. Symmetries that vary with location are central to the modern description of physical interactions with the help of gauge theory.

Generalisations

Group-like Structures				
	Totality	***Associativity***	***Identity***	***Inverses***
Group	Yes	Yes	Yes	Yes
Monoid	Yes	Yes	Yes	No
Semigroup	Yes	Yes	No	No
Loop	Yes	No	Yes	Yes
Quasigroup	Yes	No	No	No
Magma	Yes	No	No	No
Groupoid	No	Yes	Yes	Yes
Category	No	Yes	Yes	No
Semicategory	No	Yes	No	No

In abstract algebra, more general structures are defined by relaxing some of the axioms defining a group. For example, if the requirement that every element has an inverse is eliminated, the resulting algebraic structure is called a monoid. The natural numbers N(including 0) under addition form a monoid, as do the nonzero integers under multiplication(Z \ {0}, ·). There is a general method to formally add inverses to elements to any (abelian) monoid, much the same way as (Q \ {0}, ·) is derived from (Z \ {0}, ·), known as the Grothendieck group. Groupoids are similar to groups except that the composition $a \cdot b$ need not be defined for all a and b. They arise in the study of more complicated forms of symmetry, often in topological and analytical structures, such as the fundamental groupoid or stacks. Finally, it is possible to generalise any of these concepts by replacing the binary operation with an arbitrary n-ary one (i.e. an operation taking n arguments). With the proper generalisation of the group axioms this gives rise to an n-ary group. The table gives a list of several structures generalising groups.

Ring (Mathematics)

In mathematics, a ring is an algebraic structure consisting of a set together with two binary operations usually called addition and multiplication, where the set is an abelian group under addition (called the additive group of the ring) and a semigroup under multiplication such that multiplication distributes over addition.[a][›] In other words the ring axioms require that addition is commutative, addition and multiplication are associative, multiplication distributes over addition, each element in the set has an additive inverse, and there exists an additive identity. One of the most common examples of a ring is the set of integers endowed with its natural operations of addition and multiplication. Certain variations of the definition of a ring are sometimes employed, and these are outlined later in the article.

The branch of mathematics that studies rings is known as ring theory. Ring theorists study properties common to both familiar mathematical structures such as integers and polynomials, and to the many less well-known mathematical structures that also satisfy the axioms of ring theory. The ubiquity of rings makes them a central organising principle of contemporary mathematics.

Ring theory may be used to understand fundamental physical laws, such as those underlying special relativity and symmetry phenomena in molecular chemistry.

The concept of a ring first arose from attempts to prove Fermat's last theorem, starting with Richard Dedekind in the 1880s. After contributions from other fields, mainly number theory, the ring notion was generalised and firmly established during the 1920s by Emmy Noether and Wolfgang Krull. Modern ring theory—a very active mathematical discipline—studies rings in their own right. To explore rings, mathematicians have devised various notions to break rings into smaller, better-understandable pieces, such as ideals, quotient rings and simple rings. In addition to these abstract properties, ring theorists also make various distinctions between the theory of commutative rings and noncommutative rings—the former belonging to algebraic number theory and algebraic geometry. A particularly rich theory has been developed for a certain special class of commutative rings, known as fields, which lies within the realm of field theory. Likewise, the corresponding theory for noncommutative rings, that of noncommutative division rings, constitutes an active research interest for noncommutative ring theorists. Since the discovery of a mysterious connection between noncommutative ring theory and geometry during the 1980s by Alain Connes, noncommutative geometry has become a particularly active discipline in ring theory.

Definition and Illustration

First Example: the Integers: The most familiar example of a ring is the set of allintegers, Z, consisting of the numbers

..., -4, -3, -2, -1, 0, 1, 2, 3, 4, ...

together with the usual operations of addition and multiplication. These operations satisfy the following properties:

- The integers form an abelian group under addition; that is:
  - o Closure axiom for addition: Given two integers a and b, their sum, $a + b$ is also an integer.
  - o Associativity of addition: For any integers, a, b and c, $(a + b) + c = a + (b + c)$. So, adding b to a, and then adding c to this result, is the same as adding c to b, and then adding this result to a.
  - o Existence of additive identity: For any integer a, $a + 0 = 0 + a = a$. Zero is called the identity element of the integers because adding 0 to any integer (in any order) returns the same integer.
  - o Existence of additive inverse: For any integer a, there exists an integer denoted by $-a$ such that $a + (-a) = (-a) + a = 0$. The element, $-a$, is called the *additive inverse* of a because adding a to $-a$ (in any order) returns the identity.
  - o Commutativity of addition: For any two integers a and b, $a + b = b + a$. So the order in which two integers are added is irrelevant.
- The integers form a multiplicative monoid (a monoid under multiplication); that is:
  - o Closure axiom for multiplication: Given two integers a and b, their product, $a \cdot b$ is also an integer.
  - o Associativity of multiplication: Given any integers, a, b and c, $(a \cdot b) \cdot c = a \cdot (b \cdot c)$. So multiplying b with a, and then multiplying c to this result, is the same as multiplying c with b, and then multiplying a to this result.
  - o Existence of multiplicative identity: For any integer a, $a \cdot 1 = 1 \cdot a = a$. So multiplying any integer with 1 (in any order) gives back that integer. One is therefore called the multiplicative identity.
- Multiplication is distributive over addition : These two structures on the integers (addition and multiplication) are compatible in the sense that
  - o $a \cdot (b + c) = (a \cdot b) + (a \cdot c)$, and
  - o $(a + b) \cdot c = (a \cdot c) + (b \cdot c)$

for any three integers a, b and c.

Formal Definition

There are some differences in exactly what axioms are used to define a ring. Here one set of axioms for a ring with identity is given, and comments on variations follow.

A ring is a set R equipped with two binary operations $+ : R \times R \to R$ and $\cdot : R \times R \to R$ (where $\times$ denotes the Cartesian product), called *addition* and *multiplication*. To qualify as a ring, the set and two operations, $(R, +, \cdot)$, must satisfy the following requirements known as the *ring axioms*.

- $(R, +)$ is required to be an *abelian group* under addition:

1. Closure under addition.
   For all a, b in R, the result of the operation $a + b$ is also in R.[c[›]]
2. Associativity of addition.
   For all a, b, c in R, the equation $(a + b) + c = a + (b + c)$ holds.
3. Existence of additive identity.
   There exists an element 0 in R, such that for all elements a in R, the equation $0 + a = a + 0 = a$ holds.
4. Existence of additive inverse. For each a in R, there exists an element b in R such that $a + b = b + a = 0$
5. Commutativity of addition. For all a, b in R, the equation $a + b = b + a$ holds.

- $(R, \cdot)$ is required to be a monoid under multiplication:

1. Closure under multiplication. For all a, b in R, the result of the operation $a \cdot b$ is also in R.[c[›]]
2. Associativity of multiplication. For all a, b, c in R, the equation $(a \cdot b) \cdot c = a \cdot (b \cdot c)$ holds.
3. Existence of multiplicative identity.[a[›]] There exists an element 1 in R, such that for all elements a in R, the equation $1 \cdot a = a \cdot 1 = a$ holds.

- The distributive laws:

1. For all a, b and c in R, the equation $a \cdot (b + c) = (a \cdot b) + (a \cdot c)$ holds.
2. For all a, b and c in R, the equation $(a + b) \cdot c = (a \cdot c) + (b \cdot c)$ holds.

This definition assumes that a binary operation on R is a function defined on $R \times R$ with values in R. Therefore, for any a and b in R, the addition $a + b$ and the product $a \cdot b$ are elements of R.

The most familiar example of a ring is the set of all integers, Z = {..., -4, -3, -2, -1, 0, 1, 2, 3, 4, ... }, together with the usual operations of addition and multiplication.

Another familiar example is the set of real numbers R, equipped with the usual addition and multiplication.

Another example of a ring is the set of all square matrices of a fixed size, with real elements, using the matrix addition and multiplication of linear algebra. In this case, the ring elements 0 and 1 are the zero matrix (with all entries equal to 0) and the identity matrix, respectively.

Notes on the Definition

In axiomatic theories, different authors sometimes use different axioms. In the case of ring theory, some authors include the axiom $1 \neq 0$ (that is, that the multiplicative identity of the ring must be different from the additive identity). In particular they do not consider the trivial ring to be a ring.

A more significant disagreement is that some authors omit the requirement of a multiplicative identity in a ring. This allows the even integers, for example, to be considered a ring, with the natural operations of addition and multiplication, because they satisfy all of the ring axioms except for the existence of a multiplicative identity. Rings that satisfy the ring axioms as given above except the axiom of multiplicative identity are sometimes called pseudo-rings. The term ring (jocular; ring without the multiplicative identity i) is also used for such rings. Rings which *do have* multiplicative identities, (and thus satisfy all of the axioms above) are sometimes for emphasis referred to as unital rings, unitary rings, rings with unity, rings with identity or rings with 1. Note that one can always embed a non-unitary ring inside a unitary ring.

There are still other more significant differences in the way some authors define a ring. For instance, some authors omit associativity of multiplication in the set of ring axioms; rings that are nonassociative are called nonassociative rings. In this article, all rings are assumed to satisfy the axioms as given above unless stated otherwise.

Second Example: the Ring Z_4

Consider the set Z_4 consisting of the numbers 0, 1, 2, 3 where addition and multiplication are defined as follows (note that for any integer x, x mod 4 is defined to be the remainder when x is divided by 4):

- For any x, y in Z_4, $x + y$ is defined to be their sum in Z (the set of *all integers*) mod 4. So we can represent the additive structure of Z_4 by the left-most table as shown.
- For any x, y in Z_4, $x \cdot y$ is defined to be their product in Z (the set of *all integers*) mod 4. So we can represent the multiplicative structure of Z_4 by the right-most table as shown.

·	**0**	**1**	**2**	**3**
0	0	0	0	0
1	0	1	2	3
2	0	2	0	2
3	0	3	2	1

·	**0**	**1**	**2**	**3**
0	0	0	0	0
1	0	1	2	3
2	0	2	0	2
3	0	3	2	1

It is simple (but tedious) to verify that Z_4 is a ring under these operations. First of all, one can use the left-most table to show that Z_4 is closed under addition (any result is either 0, 1, 2 or 3). Associativity of addition in Z_4 follows from associativity of addition in the set of *all integers*. The additive identity is 0 as can be verified by looking at the left-most table. Given an integer x, there is always an inverse of x; this inverse is given by 4 - x as one can verify from the additive table. Therefore, Z_4 is an abelian group under addition.

Similarly, Z_4 is closed under multiplication as the right-most table shows (any result above is either 0, 1, 2 or 3). Associativity of multiplication in Z_4 follows from associativity of multiplication in the set of *all integers*. The multiplicative identity is 1 as can be verified by looking at the right-most table. Therefore, Z_4 is a monoid under multiplication.

Distributivity of the two operations over each other follow from distributivity of addition over multiplication (and vice-versa) in Z (the set of *all integers*).

Therefore, this set does indeed form a ring under the given operations of addition and multiplication.

Properties of this Ring

- In general, given any two integers, x and y, if $x \,.\, y = 0$, then either x is 0 or y is 0. It is interesting to note that this does not hold for the ring (Z_4, +, .):

2 Å" 2 = 0

although neither factor is 0. In general, a *non-zero element a* of a ring, (R, +, .) is said to be a zero divisor in (R, +, .), if there exists a non-zero element b of R such that a Å" b = 0. So in this ring, the only zero divisor is 2 (note that 0 . a = 0 for *any a* in a ring (R,+, .) so 0 is *not* considered to be a zero divisor).

- A commutative ring which has no zero divisors is called an integral domain. So Z, the ring of all integers, is an integral domain (and therefore a ring), although Z_4 (the above example) *does not* form an integral domain (but is still a ring). So in general, every integral domain is a ring but not every ring is an integral domain.

Third Example: the Trivial Ring

If we define on the singleton set {0}:

$0 + 0 = 0$

$0 \times 0 = 0$

then one can verify that ({0}, +, ×) forms a ring known as the trivial ring. Since there can be only one result for any product or sum (0), this ring is both closed and associative for addition and multiplication, and furthermore satisfies the distributive law. The additive and multiplicative identities are both equal to 0. Similarly, the additive inverse of 0 is 0. The trivial ring is also a (rather trivial) example of a zero ring.

Basic Concepts

Subring

Informally, a subring of a ring is another ring that uses the "same" operations and is contained in it. More formally, suppose (R, +, ·) is a ring, and S is a subset of R such that

- for every a, b in S, $a + b$ is in S;
- for every a, b in S, $a \cdot b$ is in S;
- for every a in S, the additive inverse $-a$ of a is in S; and
- the multiplicative identity '1' of R is in S.

Let '$+_S$' and '$\cdot_S$' denote the operations '+' and '$\cdot$', restricted to $S \times S$. Then $(S, +_S, \cdot_S)$ is a subring of $(R, +, \cdot)$. Since the restricted operations are completely determined by S and the original ones, the subring is often written simply as $(S, +, \cdot)$.

For example, a subring of the complex number ring C is any subset of C that includes 1 and is closed under addition, multiplication, and negation, such as:

- The rational numbers Q
- The algebraic numbers A
- The real numbers R

If A is a subring of R, and B is a subset of A such that B is also a subring of R, then B is a subring of A.

Homomorphism

A homomorphism from a ring $(R, +, \cdot)$ to a ring $(S, \ddagger, *)$ is a function f from R to S that commutes with the ring operations; namely, such that, for all a, b in R the following identities hold:

- $f(a + b) = f(a) \ddagger f(b)$
- $f(a \cdot b) = f(a) * f(b)$

Moreover, the function f must take the identity element 1_R of '$\cdot$' to the identity element 1_S of '*'. (this is not required if identity is not required). For example, the function that maps each integer x to its remainder modulo 4 (a number in {0, 1, 2, 3}) is a homomorphism from the ring Z to the ring Z_4.

A ring homomorphism is said to be an isomorphism if there exists an inverse homomorphism to f(i.e., a ring homomorphism which is an inverse function). Equivalently, any bijective ring homomorphism is a ring isomorphism.

Ideal

The purpose of an ideal in a ring is to somehow allow one to define the quotient ring of a ring. An ideal in a ring can therefore be thought of as a generalisation of a normal subgroup in a group. More formally, let $(R, +, \cdot)$ be a ring. A subset I of R is said to be a right ideal in R if:

- $(I, +)$ is a subgroup of the underlying additive group in $(R, +, \cdot)$ (i.e. $(I, +)$ is a subgroup of $(R, +)$).
- For every x in I and r in R, $x \cdot r$ is in I.

A left ideal is similarly defined with the second condition being replaced. More specifically, a subset I of R is a left ideal in R if:

- $(I, +)$ is a subgroup of the underlying additive group in $(R, +, \cdot)$ (i.e. $(I, +)$ is a subgroup of $(R, +)$).
- For every x in I and r in R, $r \cdot x$ is in I.

Notes

- If k is in R, then the set of elements $k \cdot R$ is a right ideal in R, and the set of elements $R \cdot k$ is a left ideal in R. These ideals (for any k in R) are called the principal right and left ideals generated by k.
- If every ideal in a ring $(R, +, \cdot)$ is a principal ideal in $(R, +, \cdot)$, $(R, +, \cdot)$ is said to be a principal ideal ring.
- An ideal in a ring, $(R, +, \cdot)$, is said to be a *two-sided ideal* if it is both a left ideal and right ideal in $(R, +, \cdot)$. It is preferred to call a two-sided ideal, simply an *ideal*.
- If $I = \{0\}$ (where 0 is the additive identity of the ring $(R, +, \cdot)$), then I is an ideal known as the *trivial ideal*. Similarly, R is also an ideal in $(R, +, \cdot)$ called the *unit ideal*.

Examples

- Any additive subgroup of the integers is an ideal in the integers with its natural ring structure.
- There are no non-trivial ideals in R (the ring of all real numbers) (i.e., the only ideals in R are {0} and R itself). More generally, a field cannot contain any non-trivial ideals.
- From the previous example, every field must be a principal ideal ring.
- A subset, I, of a *commutative ring* $(R, +, \cdot)$ is a left ideal if and only if it is a right ideal. So for simplicity's sake, we refer to any ideal in a commutative ring as just an *ideal*.

History

The study of rings originated from the theory of polynomial rings and the theory of algebraic integers. Furthermore, the appearance of hypercomplex numbers in the mid-19th century undercut the pre-eminence of fields in mathematical analysis.

In the 1880s Richard Dedekind introduced the concept of a ring, and the termring (Zahlring) was coined by David Hilbert in 1892 and published in the article Die Theorie der algebraischen Zahlkφrper, Jahresbericht der Deutschen Mathematiker Vereinigung,

Vol. 4, 1897. According to Harvey Cohn, Hilbert used the term for a specific ring that had the property of "circling directly back" to an element of itself.

The first axiomatic definition of a ring was given by Adolf Fraenkel in an essay in Journal für die reine und angewandte Mathematik (A. L. Crelle), vol. 145, 1914. In 1921, Emmy Noether gave the first axiomatic foundation of the theory of commutative rings in her monumental paper Ideal Theory in Rings.

Elementary Properties of Rings

Some properties of rings follow directly from the ring axioms through very simple proofs.

In particular, since the axioms state that (R,+) is a commutative group, all pertinent theorems of group theory apply, such as the uniqueness of the additive identity and the uniqueness of the additive inverse of a particular element. In the same way one can prove the uniqueness of the inverse of a unit in a ring.

However, rings also have specific properties that combine addition with multiplication. In any ring (R, +, .):

- For any element a, the equations $a \,.\, 0 = 0 \,.\, a = 0$ hold.
- If $0 = 1$, the ring is trivial (that is, R has only one element).
- For any element a, the equation $(-1) \,.\, a = -a$ holds.
- For any elements a and b, the equations $(-a) \,.\, b = a \,.\, (-b) = -(a \,.\, b)$ hold.

New Rings from Old

Quotient Ring: Informally, the quotient ring of a ring, is a generalisation of the notion of a quotient group of a group. More formally, given a ring $(R, +, \cdot)$ and a two-sided ideal I of $(R, +, \cdot)$, the quotient ring (or factor ring) R/I is the set of cosets of I (with respect to the underlying additive group of $(R, +, \cdot)$; i.e. cosets with respect to $(R, +)$) together with the operations:

$(a + I) + (b + I) = (a + b) + I$ and

$(a + I)(b + I) = (ab) + I.$

For every a, b in R.

Polynomial Ring

Formal Definition: Let $(R, +_R, \cdot_R)$ be a ring and let $S = \{(f_i)_{i \in N} : f_i \in R \text{ and } f_i = 0 \text{ for all but finitely many } i \in \mathbb{N}\}$ where the convention that $\mathbb{N}$ is the set of all nonnegative integers, is adopted. Define $+_S : S \times S \to S$ and $\cdot_S : S \times S \to S$, as follows (where $a = (a_i)_{i\in\mathbb{N}}$ and $b = (b_i)_{i\in\mathbb{N}}$ are arbitrary elements of S):

$$a +_S b = (a_i +_R b_i)_{i\in\mathbb{N}}$$

$$a \cdot_S b = (\sum_{j=0}^{i} a_j \cdot_R b_{i-j})_{i\in\mathbb{N}}$$

Then $(S, +_S, \cdot_S)$ is a ring referred to as the polynomial ring over R.

Notes

Most authors write S as $R[X]$, where $X = (0,1,0,0,0\ldots)$ is an element of S. This convention is adopted because if $f = (f_i)_{i\in\mathbb{N}}$ is an element of S, and if $C = \{i \in \mathbb{N} : f_i \neq 0\}$, f can be written as a polynomial with coefficients in R, in the variable X, as follows:

$$f = \sum_{c\in C} f_c \cdot_S X^c$$

This allows one to view S as merely the set of all polynomials over R in the variable X, with multiplication and addition of polynomials defined in the canonical manner. Therefore, in all that follows, S, denoted by $R[X]$, shall be identified in this fashion. Addition and multiplication in S, and those of the underlying ring R, will be denoted by juxtaposition.

Matrix Ring

Formal Definition: Let $(R,+_R, \cdot_R)$ be a ring and let $M_r(R) = \{(f_{ij})_{i,j\in\mathbb{N}_r} : f_{ij} \in R \text{ for every } i,j \in \{1,2,3,\ldots,r\}\}$. Define $+_M : M_r(R) \times M_r(R) \to M_r(R)$ and $\cdot_M : M_r(R) \times M_r(R) \to M_r(R)$, as follows (where $a = (a_{ij})_{i,j\in\mathbb{N}_r}$ and $b = (b_{ij})_{i,j\in\mathbb{N}_r}$ are arbitrary elements of $M_r(R)$):

$$a +_M b = (a_{ij} +_R b_{ij})_{i,j\in\mathbb{N}_r}$$

$$a \cdot_M b = (\sum_{k=1}^{r} a_{ik} \cdot_R b_{kj})_{i,j\in\mathbb{N}_r}$$

Then $(M_r(R), +_M, \cdot_M)$ is a ring referred to as the ring of $r\times r$ matrices over R.

Notes

The ring $M_r(R)$ can be viewed as merely the ring of $r\times r$ matrices over R with matrix addition and multiplication defined in the canonical manner. For given $f = (f_{ij})_{i,j\in\mathbb{N}_r}$ in $M_r(R)$, f can be identified with the $r \times r$ matrix whose (i, j)-entry is f_{ij}. Therefore, in all that follows, $M_r(R)$ and each of its elements shall be identified in this fashion. Addition and multiplication in $M_r(R)$, and those of the underlying ring R, will be denoted by juxtaposition.

Some Examples of the Ubiquity of Rings

Many different kinds of mathematical objects can be fruitfully analysed in terms of some associated ring.

Cohomology Ring of a Topological Space

To any topological space X one can associate its integral cohomology ring

$$H^*(X,\mathbb{Z}) = \bigoplus_{i=0}^{\infty} H^i(X,\mathbb{Z}),$$

a graded ring. There are also homology groups $H_i(X,\mathbb{Z})$ of a space, and indeed these were defined first, as a useful tool for distinguishing between certain pairs of topological spaces, like the spheres and tori, for which the methods of point-set topology are not well-suited. Cohomology groups were later defined in terms of homology groups in a way which is roughly analogous to the dual of a vector space. To know each individual integral homology group is essentially the same as knowing each individual integral cohomology group, because of the universal coefficient theorem. However, the advantage of the cohomology groups is that there is a natural product, which is analogous to the observation that one can multiply pointwise a k-multilinear form and an l-multilinear form to get a $(k+l)$-mulilinear form.

The ring structure in cohomology provides the foundation for characteristic classes of fibre bundles, intersection theory on manifolds and algebraic varieties, Schubert calculus and much more.

Burnside Ring of a Group

To any group is associated its Burnside ring which uses a ring to describe the various ways the group can act on a finite set. The Burnside ring's additive group is the free abelian group whose basis are the transitive actions of the group and whose addition is the disjoint union of the action. Expressing an action in terms of the basis is decomposing an action into its transitive constituents. The multiplication is easily expressed in terms of the representation ring: the multiplication in the Burnside ring is formed by writing the tensor product of two permutation modules as a permutation module. The ring structure allows a formal way of subtracting one action from another. Since the

Burnside ring is contained as a finite index subring of the representation ring, one can pass easily from one to the other by extending the coefficients from integers to the rational numbers.

Representation Ring of a Group Ring

To any group ring or Hopf algebra is associated its representation ring or "Green ring". The representation ring's additive group is the free abelian group whose basis are the indecomposable modules and whose addition corresponds to the direct sum. Expressing a module in terms of the basis is finding an indecomposable decomposition of the module. The multiplication is the tensor product. When the algebra is semisimple, the representation ring is just the character ring from character theory, which is more or less the Grothendieck group given a ring structure.

Function Field of an Irreducible Algebraic Variety

To any irreducible algebraic variety is associated its function field. The points of an algebraic variety correspond to valuation rings contained in the function field and containing the coordinate ring. The study of algebraic geometry makes heavy use of commutative algebra to study geometric concepts in terms of ring-theoretic properties. Birational geometry studies maps between the subrings of the function field.

Face Ring of a Simplicial Complex

Every simplicial complex has an associated face ring, also called its Stanley–Reisner ring. This ring reflects many of the combinatorial properties of the simplicial complex, so it is of particular interest in algebraic combinatorics. In particular, the algebraic geometry of the Stanley–Reisner ring was used to characterise the numbers of faces in each dimension of simplicial polytopes.

Special Rings and Rings with Additional Structure

Finite Ring

Given a natural number m, how many distinct rings (not necessarily with unity) have m elements? When m is prime there are only two rings of order m; the additive group of each being isomorphic to the cyclic group of order m. One is the zero ring and the other is the Galois field.

As with finite groups, the complexity of the classification depends upon the complexity of the prime factorisation of m. If m is the square of a prime, for instance, there are precisely eleven rings having order *m*. On the other hand, there can be only two *groups* having order *m*; both of which are abelian.

The theory of finite rings is more complex than that of finite abelian groups, since any finite abelian group is the additive group of at least two nonisomorphic finite rings: the direct product of copies of $\mathbb{Z}/m\mathbb{Z}$, and the zero ring. On the other hand, the theory of finite rings is simpler than that of not necessarily abelian finite groups. For instance, the classification of finite simple groups was one of the major breakthroughs of 20th century mathematics, its proof spanning thousands of journal pages. On the other hand, any finite simple ring is isomorphic to the ring $M_n(\mathbb{F}_q)$ of n by n matrices over a finite field of order q. This follows from two theorems of Joseph Wedderburn established in 1905 and 1907. One of these theorems, known as Wedderburn's little theorem, asserts that any finite division ring is necessarily commutative. Nathan Jacobson later discovered yet another condition which guarantees commutativity of a ring:

If for every element r of R there exists an integer n > 1 such that $r^n = r$, then R is commutative.

If, $r^2 = r$ for every *r*, the ring is called a Boolean ring. More general conditions which guarantee commutativity of a ring are also known.

The number of rings with *m* elements, for *m* a natural number, is listed under A027623 in the On-Line Encyclopedia of Integer Sequences.

Associative Algebras

An associative algebra is a ring that is also a vector space over a field *K*. For instance, the set of *n* by *n* matrices over the real field R has dimension n^2 as a real vector space, and the matrix multiplication corresponds to the ring multiplication. For a non-trivial but elementary example consider 2 × 2 real matrices.

Lie Ring

A Lie ring is defined to be a ring that is nonassociative and anticommutative under multiplication, that also satisfies the Jacobi identity. More specifically we can define a Lie ring *L* to be an abelian groupunder addition with an operation $[\cdot,\cdot]$ that has the following properties:

- *Bilinearity*:

$$[x+y,z]=[x,z]+[y,z], \quad [z,x+y]=[z,x]+[z,y]$$

for all $x, y, z \in L$.

- *Jacobi identity*:

$$[x,[y,z]]+[y,[z,x]]+[z,[x,y]]=0$$

for all x, y, z in L.

- For all x in L.

$$[x,x]=0$$

Lie rings need not be Lie groups under addition. Any Lie algebra is an example of a Lie ring. Any associative ring can be made into a Lie ring by defining a bracket operator $[x,y] = xy - yx$. Conversely to any Lie algebra there is a corresponding ring, called the universal enveloping algebra.

Lie rings are used in the study of finite p-groups through the Lazard correspondence. The lower central factors of a p-group are finite abelian p-groups, so modules over Z/pZ. The direct sum of the lower central factors is given the structure of a Lie ring by defining the bracket to be the commutator of two coset representatives. The Lie ring structure is enriched with another module homomorphism, then pth power map, making the associated Lie ring a so-called restricted Lie ring.

Lie rings are also useful in the definition of a p-adic analytic groups and their endomorphisms by studying Lie algebras over rings of integers such as the p-adic integers. The definition of finite groups of Lie type due to Chevalley involves restricting from a Lie algebra over the complex numbers to a Lie algebra over the integers, and the reducing modulo p to get a Lie algebra over a finite field.

Topological Ring

Let (X, T) is a topological space and $(X, +, \cdot)$ be a ring. Then $(X, T, +, \cdot)$ is said to be a topological ring, if its ring structure and topological structure are both compatible (i.e work together) over each other. That is, the addition map $(+ : X\times X \to X)$ and the multiplication map $(\cdot : X\times X \to X)$ have to be both continuous as maps between topological spaces where X x Xinherits the product topology. So clearly, any topological ring is a topological group (under addition).

Examples

- The set of all real numbers, R, with its natural ring structure and the standard topology forms a topological ring.

- The direct product of two topological rings is also a topological ring.

Commutative Rings

Although ring addition is commutative, so that for every a, b in R, $a + b = b + a$, ring multiplication is not required to be commutative; $a \cdot b$ need not equal $b \cdot a$ for all a, b in R. Rings that also satisfy commutativity for multiplication are called commutative rings.[e][›] Formally,

Formal Definition

Let $(R, +, \cdot)$ be a ring. Then $(R, +, \cdot)$ is said to be a commutative ring if for every a, b in R, $a \cdot b = b \cdot a$. That is, $(R, +, \cdot)$ is required to be a commutative monoid under multiplication.

Examples

- The integers form a commutative ring under the natural operations of addition and multiplication.
- An example of a noncommutative ring is the ring of $n \times n$ matrices over a non-trivial field K, for$n > 1$. In particular, the ring of all 2×2 matrices over R (the set of all real numbers) do not form a commutative ring as the following computation shows:

$$\begin{bmatrix} 1 & 1 \\ 0 & 1 \end{bmatrix} \cdot \begin{bmatrix} 1 & 1 \\ 1 & 0 \end{bmatrix} = \begin{bmatrix} 2 & 1 \\ 1 & 0 \end{bmatrix}, \text{which is not equal to} \begin{bmatrix} 1 & 1 \\ 1 & 0 \end{bmatrix} \cdot \begin{bmatrix} 1 & 1 \\ 0 & 1 \end{bmatrix} = \begin{bmatrix} 1 & 2 \\ 1 & 1 \end{bmatrix}.$$

Principal Ideal Domains

Although rings are structurally similar to the integers, there are certain ring-theoretic properties that the integers may satisfy but a general ring may not. One such property is the requirement that every ideal in a ring be generated by a single element; that is, be a principal ideal. Formally,

Definition

Let R be a ring. Then R is said to be a principal ideal ring (abbreviated PIR), if every ideal in R is of the form $a \cdot R = \{a \cdot r \mid r \text{ in } R\}$. A principal ideal domain (PID) is a principal ideal ring that is also an integral domain.

The requirement that a ring be a principal ideal domain is somewhat stronger than the other more common properties a ring may satisfy. For example, it is true that if a ring, R, is a unique factorisation domain (UFD), then the polynomial ring over R is also a UFD. However, such a result does not in general hold for principal ideal rings. For example, the integers are an easy example of a principal ideal ring, but the polynomial ring over the integers fails to be a PIR; if $R = Z[x]$ denotes the polynomial ring over the integers, $I = 2 \cdot R + X \cdot R$ is an ideal which cannot be generated by a single element. Despite this counter example, the polynomial ring over any field is always a principal ideal domain and in fact, a Euclidean domain. More generally, a polynomial ring is a PID if and only if the polynomial ring in question is over a field.

Aside from the polynomial ring over a PIR, principal ideal rings possess many interesting properties due to their connection with the integers in terms of divisibility; that is, principal ideal domains behave similarly to the integers with respect to divisibility. For example, any PID is a UFD; i.e., an analogue of the fundamental theorem of arithmetic holds for principal ideal domains.

Furthermore, since Noetherian rings are precisely those rings in which any ideal is finitely generated, principal ideal domains are trivially Noetherian rings. The fact that irreducible elements coincide with prime elements for PID's, together with the fact that every PID is Noetherian, implies that any PID is a UFD. One can also speak of the greatest common divisor of two elements in a PID; if x and y are elements of R, a principal ideal domain, then $x \cdot R + y \cdot R = c \cdot R$ for some c in R since the left-hand side is indeed an ideal. Therefore, c is the desired "GCD" of x and y.

An important class of rings lying in between fields and PID's, is the class of Euclidean domains. In particular, any field is a Euclidean domain, and any Euclidean domain is a PID. An ideal in a Euclidean domain is generated by any element of that ideal with minimum degree (all such elements must be associate). However, not every PID is a Euclidean domain; the ring $\mathbb{Z}\left[\frac{1+\sqrt{-19}}{2}\right]$ furnishes a counter example.

Unique Factorisation Domains

The theory of unique factorisation domains (UFD) also forms an important part of ring theory. In effect, a unique factorisation domain is ring in which an analogue of the fundamental theorem of arithmeticholds. Formally,

Definition

Let R be a ring. Then R is said to be a unique factorisation domain (abbreviated UFD), if the following conditions are satisfied:

1. R is an integral domain.
2. Every non-zero non-unit of R is the product of a finite number of irreducible elements.
3. If $\prod_{i=1}^{n} a_i = \prod_{j=1}^{m} b_j$, where all a_i's and b_j's are irreducible, then $n = m$ and after possible renumbering of the a_i's and b_j's, $b_i = a_i \cdot u_i$ where u_i is a unit in R.

The second condition above guarantees that "non-trivial" elements of R can be decomposed into irreducibles, and according to the third condition, such a finite decomposition is unique "up to multiplication by unit elements." This weakened form of uniqueness is reasonable to assume for otherwise even the integers would not satisfy the properties of being a UFD ($(-2)^2 = 2^2 = 4$ demonstrates two "distinct" decompositions of 4; however both decompositions of 4 are equivalent up to multiplication by units (-1 and +1)). The fact that the integers constitute a UFD follows from the fundamental theorem of arithmetic.

For arbitrary rings, one may define a prime element and an irreducible element; these two may not in general coincide. However, a prime element in a domain is always irreducible. For UFD's, irreducible elements are also primes.

The class of unique factorisation domains is related to other classes of rings. For instance, any Noetherian domain satisfies conditions 1 and 2 above, but in general Noetherian domains fail to satisfy condition 3. However, if the set of prime elements and the set of irreducible elements coincide for a Noetherian domain, the third condition of a UFD is satisfied. In particular, principal ideal domains are UFD's.

Integral Domains and Fields

While rings are very important mathematical objects, there are a lot of restrictions involved in their theory. For instance, suppose we have a ring R and suppose a and b are in R. If a is non-zero and $a \cdot b = 0$, then b need not necessarily be 0. In particular, if $a \cdot b = a \cdot c$ with a non-zero b need not equal c. An example of this is the set of n x n matrices over R where a maybe such a non-zero matrix but may be singular. In this case, the result may not be true. However, we can impose additional conditions on the ring to ensure that this be true; namely make the ring into an integral domain (that is a non-trivial commutative ring with no zero divisors).

But we still run into a problem; namely we can't necessarily divide by non-zero elements. For example, the collection of all integers form an integral domain but we still can't divide an integer a by another integer b. For example, 2 cannot divide 3 to obtain another element in this ring.

However, this problem can readily be solved if we ensure that every element in the ring has a multiplicative inverse. A field is a ring in which the non-zero elements form an abelian group under multiplication. In particular, a field is an integral domain (and therefore has no zero divisors) along with an additional operation of 'division'. Namely, if a and b are in a field F, then a/b is defined to be $a \cdot b^{-1}$ which is well defined.

Formal Definition

Let $(R, +, \cdot)$ be a ring. Then $(R, +, \cdot)$ is said to be an integral domain if $(R, +, \cdot)$ is commutative and has no zero divisors. Furthermore, $(R, +, \cdot)$ is said to be a field, if its non-zero elements form an abelian group under multiplication.

Note

- If we require that the *zero element* (0) in a ring to have a multiplicative inverse, then the ring must be trivial.

Examples

- The integers form a commutative ring under the natural operations of addition and multiplication. In fact, the integers form what is known as an integral domain (a commutative ring with no zero divisors).
- A ring whose non-zero elements form an abelian group under multiplication (not just a commutative monoid), is called a field. So every field is an integral domain and every integral domain is a commutative ring. Furthermore, any finite integral domain is a field.

Relation to Other Algebraic Structures

The following is a chain of class inclusions that describes the relationship between rings, domains and fields:

- Commutative rings ⊃ integral domains ⊃ half factorisation domains ⊃ unique factorisation domains ⊃ principal ideal domains ⊃ Euclidean domains ⊃ fields

Fields and integral domains are very important in modern algebra.

Noncommutative Rings

The study of noncommutative rings is a major area in modern algebra; especially ring theory. Often noncommutative rings possess interesting invariants that commutative rings do not. As an example, there exist rings which contain non-trivial proper left or right ideals, but are still simple; that is contain no non-trivial proper (two-sided) ideals. This example illustrates how one must take care when studying noncommutative rings because of possible counterintuitive misconceptions.

The theory of vector spaces is one illustration of a special case of an object studied in noncommutative ring theory. In linear algebra, the "scalars of a vector space" are required to lie in a field - a commutative division ring. The concept of a module, however, requires only that the scalars lie in an abstract ring. Neither commutativity nor the division ring assumption is required on the scalars in this case. Module theory has various applications in noncommutative ring theory, as one can often obtain information about the structure of a ring by making use of its modules.

The concept of the Jacobson radical of a ring; that is, the intersection of all right/left annihilators of simple right/left modules over a ring, is one example. The fact that the Jacobson radical can be viewed as the intersection of all maximal right/left ideals in the ring, shows how the internal structure of the ring is reflected by its modules. It is also remarkable that the intersection of all maximal right ideals in a ring is the same as the intersection of all maximal left ideals in the ring, in the context of all rings; whether commutative or noncommutative. Therefore, the Jacobson radical also captures a concept which may seem to be not well-defined for noncommutative rings.

Noncommutative rings serve as an active area of research due to their ubiquity in mathematics. For instance, the ring of n by n matrices over a field is noncommutative despite its natural occurrence in physics. More generally, endomorphism rings of abelian groups are rarely commutative.

Noncommutative rings, like noncommutative groups, are not very well understood. For instance, although every finite abelian group is the direct sum of (finite) cyclic groups of prime-power order, non-abelian groups do not possess such a simple structure. Likewise, various invariants exist for commutative rings, whereas invariants of noncommutative rings are difficult to find. As an example, the nilradical, although "innocent" in nature, need not be an ideal unless the ring is assumed to be commutative. Specifically, the set of all nilpotent elements in the ring of all n x n matrices over a division ring never forms an ideal, irrespective of the division ring chosen. Therefore, the nilradical cannot be studied in noncommutative ring theory. Note however that there are analogues of the nilradical defined for noncommutative rings, that coincide with the nilradical when commutativity is assumed.

One of the best known noncommutative rings is the division ring of quaternions.

Category Theoretical Description

Every ring can be thought of as a monoid in Ab, the category of abelian groups (thought of as amonoidal category under the tensor product of $\mathbb{Z}$-modules). The monoid action of a ring R on a abelian group is simply an R-module. Essentially, an R-module is a generalisation of the notion of a vector space - where rather than a vector space over a field, one has a "vector space over a ring".

Let (A, +) be an abelian group and let End(A) be its endomorphism ring. Note that, essentially, End(A) is the set of all morphisms of A, where if f is in End(A), and g is in End(A), the following rules may be used to compute f + g and f · g:

- $(f + g)(x) = f(x) + g(x)$
- $(f \cdot g)(x) = f(g(x))$

where + as in $f(x) + g(x)$ is addition in A, and function composition is denoted from right to left. Therefore, associated to any abelian group, is a ring. Conversely, given any ring, $(R, +, \cdot)$, $(R, +)$ is an abelian group. Furthermore, for every r in R, right (or left) multiplication by r gives rise to a morphism of $(R, +)$, by right (or left) distributivity. Let $A = (R, +)$. Consider those endomorphisms of A, that "factor through" right (or left) multiplication of R. In other words, let End $_R(A)$ be the set of all morphisms m of A, having the property that $m(r \cdot x) = r \cdot m(x)$. It was seen that every r in R gives rise to a morphism of A - right multiplication by r. It is in fact true that this association of any element of R, to a morphism of A, as a function from R to End $_R(A)$, is an isomorphism of rings. In this sense, therefore, any ring can be viewed as the endomorphism ring of some abelian X-group (by X-group, it is meant a group with *X* being its set of operators). In essence, the most general form of a ring, is the endomorphism group of some abelian X-group.

Field

In abstract algebra, a field is a commutative ring whose nonzero elements form a group under multiplication. As such it is an algebraic structure with notions of addition, subtraction, multiplication, and division, satisfying certain axioms. The most commonly used fields are the field of real numbers, the field of complex numbers, and the field of rational numbers, but there are also finite fields, fields of functions, various algebraic number fields, *p*-adic fields, and so forth. To avoid confusion with other uses of the word "field", the term "corpus" may also be used.

Any field may be used as the scalars for a vector space, which is the standard general context forlinear algebra. The theory of field extensions (including Galois theory) involves the roots of polynomials with coefficients in a field; among other results, this theory leads to impossibility proofs for the classical problems of angle trisection and squaring the circle with a compass and straight edge, as well as a proof of the Abel–Ruffini theorem on the algebraic insolubility of quintic equations. In modern mathematics, the theory of fields (or field theory) plays an essential role in number theory and algebraic geometry.

As an algebraic structure, every field is a ring, but not every ring is a field. The most important difference is that fields allow for division (though not division by zero), while a ring need not possess multiplicative inverses; for example the integers form a ring, but $2x = 1$ has no solution in integers. Also, the multiplication operation in a field is required to be commutative. A ring in which division is possible but commutativity is not assumed (such as the quaternions) is called a *division ring* or *skew field*. (Historically, division rings were sometimes referred to as fields, while fields were called commutative *fields*.)

As a ring, a field may be classified as a specific type of integral domain, and can be characterised by the following (not exhaustive) chain of class inclusions:

Commutative rings ⊃ integral domains ⊃ integrally closed domains ⊃ unique factorisation domains ⊃ principal ideal domains ⊃ Euclidean domains ⊃ fields ⊃ finite fields.

Definition and Illustration

Intuitively, a field is a set F that is a commutative group with respect to two compatible operations, addition and multiplication, with "compatible" being formalised by distributivity, and the caveat that the additive identity (0) has no multiplicative inverse (one cannot divide by 0).

The most common way to formalise this is by defining a field as a set together with two operations, usually called addition and multiplication, and denoted by + and ·, respectively, such that the following axioms hold; subtraction and division are defined implicitly in terms of the inverse operations of addition and multiplication:

Closure of F Under Addition and Multiplication

For all a, b in F, both $a + b$ and $a \cdot b$ are in F (or more formally, + and · are binary operations on F).

Associativity of Addition and Multiplication

For all a, b, and c in F, the following equalities hold: $a + (b + c) = (a + b) + c$ and $a \cdot (b \cdot c) = (a \cdot b) \cdot c$.

Commutativity of Addition and Multiplication

For all a and b in F, the following equalities hold: $a + b = b + a$ and $a \cdot b = b \cdot a$.

Existence of additive and multiplicative *identity elements*

There exists an element of F, called the *additive identity* element and denoted by 0, such that for all a in F, $a + 0 = a$. Likewise, there is an element, called the *multiplicative identity* element and denoted by 1, such that for all a in F, $a \cdot 1 = a$. To exclude the trivial ring, the additive identity and the multiplicative identity are required to be distinct.

Existence of Additive Inverses and Multiplicative Inverses

For every a in F, there exists an element $-a$ in F, such that $a + (-a) = 0$. Similarly, for any a in F other than 0, there exists an element a^{-1} in F, such that $a \cdot a^{-1} = 1$. (The elements $a + (-b)$a nd a · b^{-1} are lao denoted $a - b$ and a/b, respectively.) In other words, *subtraction* and *division operations* exist.

Distributivity of Multiplication Over Addition

For all a, b and c in F, the following equality holds: $a \cdot (b + c) = (a \cdot b) + (a \cdot c)$.

A field is therefore an algebraic structure 0F, +, ·, –, $^{-1}$, 0, 1 2, 1, 1, 0, 0□, consisting of two abelian groups:

- F under +, –, and 0;
- $F\setminus\{0\}$ under $\cdot$, $^{-1}$, and 1, with $0 \neq 1$,

with $\cdot$ distributing over +.

First Example: Rational Numbers

A simple example of a field is the field of rational numbers, consisting of the fractions a/b, where a and b are integers, and $b \neq 0$. The additive inverse of such a fraction is simply $-a/b$, and the multiplicative inverse (provided that $a \neq 0$) is b/a.

$$\frac{b}{a}\cdot\frac{a}{b}=\frac{ba}{ab}=1.$$

The abstractly required field axioms reduce to standard properties of rational numbers, such as the law of distributivity

$$\frac{a}{b}\cdot\left(\frac{c}{d}+\frac{e}{f}\right)=\frac{a}{b}\cdot\frac{cf+ed}{df}=\frac{a(cf+ed)}{bdf}=\frac{acf}{bdf}+\frac{aed}{bdf}=\frac{ac}{bd}+\frac{ae}{bf}=\frac{a}{b}\cdot\frac{c}{d}+\frac{a}{b}\cdot\frac{e}{f},$$

or the law of commutativity and law of associativity.

Second Example: a Field with Four Elements

+	O	I	A	B
O	O	I	A	B
I	I	O	B	A
A	A	B	O	I
B	B	A	I	O

·	O	I	A	B
O	O	O	O	O
I	O	I	A	B
A	O	A	B	I
B	O	B	I	A

In addition to familiar number systems such as the rationals, there are other, less immediate examples of fields. The following example is a field consisting of four elements called O, I, A and B. The notation is chosen such that O plays the role of the additive identity element (denoted 0 in the axioms), and I is the multiplicative identity (denoted 1 above). One can check that all field axioms are satisfied. For example:

A · (B + A) = A · I = A, which equals A · B + A · A = I + B = A, as required by the distributivity.

The above field is called a finite field with four elements, and can be denoted F_4. Field theory is concerned with understanding the reasons for the existence of this field, defined in a fairly ad-hoc manner, and describing its inner structure. For example, from a glance at the multiplication table, it can be seen that any non-zero element (i.e., I, A, and B) is a power of A: $A = A^1$, $B = A^2 = A \cdot A$, and finally $I = A^3 = A \cdot A \cdot A$. This is not a coincidence, but rather one of the starting points of a deeper understanding of (finite) fields.

Alternative Axiomatisations

As with other algebraic structures, there exist alternative axiomatisations. Because of the relations between the operations, one can alternatively axiomatise a field by explicitly assuming that are four binary operations (add, subtract, multiply, divide) with axioms relating these, or in terms of two binary operations (add, multiply) and two unary operations (additive inverse, multiplicative inverse), or other variants.

The usual axiomatisation in terms of the two operations of addition and multiplication is brief and allows the other operations to be defined in terms of these basic ones, but in other contexts, such as topology and category theory, it is important to include all operations as explicitly given, rather than implicitly defined (compare topological group).

This is because without further assumptions, the implicitly defined inverses may not be continuous (in topology), or may not be able to be defined (in category theory): defining an inverse requires that one be working with a set, not a more general object.

For a very economical axiomatisation of the field of real numbers, whose primitives are merely a set R with 1"R, addition, and a binary relation, <,.

Related Algebraic Structures

Ring and Field Axioms

	Abelian Group	*Ring*	*Commutative Ring*	*Skew Field or Division Ring*	*Field*
Abelian (additive) group structure	Yes	Yes	Yes	Yes	Yes
Multiplicative structure and distributivity	–	Yes	Yes	Yes	Yes
Commutativity of multiplication	–	No	Yes	No	Yes
Multiplicative inverses	–	No	No	Yes	Yes

The axioms imposed above resemble the ones familiar from other algebraic structures. For example, the existence of the binary operation "·", together with its commutativity, associativity, (multiplicative) identity element and inverses are precisely the axioms for an abelian group.

In other words, for any field, the subset of nonzero elements $F \setminus \{0\}$, also often denoted $F^{\times}$, is an abelian group ($F^{\times}$, ·) usually called multiplicative group of the field. Likewise (F, +) is an abelian group. The structure of a field is hence the same as specifying such two group structures (on the same set), obeying the distributivity.

Important other algebraic structures such as rings arise when requiring only part of the above axioms. For example, if the requirement of commutativity of the multiplication operation · is dropped, one gets structures usually called division rings or *skew fields*.

Remarks

By elementary group theory, applied to the abelian groups ($F^{\times}$, ·), and (F, +), the additive inverse "a and the multiplicative inverse a^{-1} are uniquely determined by a.

Similar direct consequences from the field axioms include

$-(a \cdot b) = (-a) \cdot b = a \cdot (-b)$, in particular $-a = (-1) \cdot a$

as well as

$$a \cdot 0 = 0.$$

Both can be shown by replacing b or c with 0 in the distributive property

History

The concept of field was used implicitly by Niels Henrik Abel and Évariste Galois in their work on the solvability of polynomial equations with rational coefficients of degree five or higher.

In 1857, Karl von Staudt published his Algebra of Throws which provided a geometric model satisfying the axioms of a field. This construction has been frequently recalled as a contribution to the foundations of mathematics.

In 1871, Richard Dedekind introduced, for a set of real or complex numbers which is closed under the four arithmetic operations, the German word Kφrper, which means "body" or "corpus" (to suggest an organically closed entity), hence the common use of the letter K to denote a field. He also defined rings (then called order or order-modul), but the term "a ring" (Zahlring) was invented by Hilbert. In 1893, Eliakim Hastings Moore called the concept "field" in English.

In 1881, Leopold Kronecker defined what he called a "domain of rationality", which is indeed a field of polynomials in modern terms. In 1893, Heinrich M. Weber gave the first clear definition of an abstract field. In 1910, Ernst Steinitz published the very influential paper Algebraische Theorie der Kφrper (English: Algebraic Theory of Fields). In this paper he axiomatically studies the properties of fields and defines many important field theoretic concepts like prime field, perfect field and the transcendence degree of a field extension.

Emil Artin developed the relationship between groups and fields in great detail from 1928 through 1942.

Examples

Rationals and Algebraic Numbers: The field of rational numbers Q has been introduced above. A related class of fields very important in number theory are algebraic number fields. We will first give an example, namely the field Q(ζ) consisting of numbers of the form

$$a + b\zeta$$

with $a, b \in Q$, where ζ is a primitive third root of unity, i.e., a complex number satisfying $\zeta^3 = 1$, $\zeta \neq 1$. This field extension can be used to prove a special case of Fermat's last theorem, which asserts the non-existence of rational nonzero solutions to the equation

$$x^3 + y^3 = z^3.$$

In the language of field extensions detailed below, Q(ζ) is a field extension of degree 2. Algebraic number fields are by definition finite field extensions of Q, that is, fields containing Q having finite dimension as a Q-vector space.

Reals, Complex Numbers, and P-adic Numbers

Take the real numbers R, under the usual operations of addition and multiplication. When the real numbers are given the usual ordering, they form a *complete ordered field*; it is this structure which provides the foundation for most formal treatments of calculus.

The complex numbers C consist of expressions

$$a + bi$$

where i is the imaginary unit, i.e., a (non-real) number satisfying $i^2 = -1$. Addition and multiplication of real numbers are defined in such a way that all field axioms hold for C. For example, the distributive law enforces

$(a + bi) \cdot (c + di) = ac + bci + adi + bdi^2$, which equals $ac - bd + (bc + ad)i$.

The real numbers can be constructed by completing the rational numbers, i.e., filling the "gaps": for example -2 is such a gap. By a formally very similar procedure, another important class of fields, the field of *p*-adic numbers Q_p is built. It is used in number theory and *p*-adic analysis.

Hyperreal numbers and superreal numbers extend the real numbers with the addition of infinitesimal and infinite numbers.

Constructible Numbers

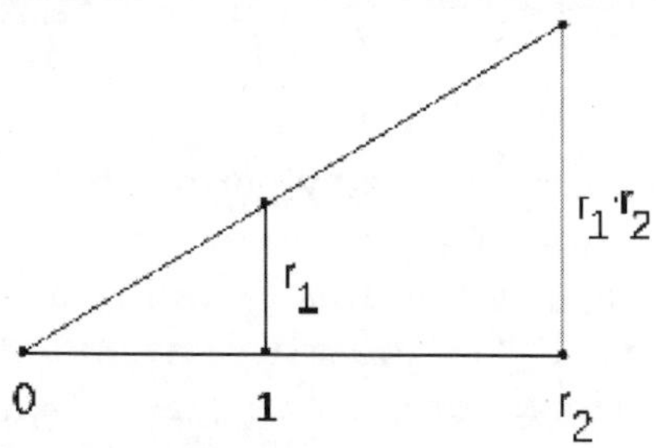

Figure 12: Given 0, 1, r_1 and r_2, the construction yields $r_1 \cdot r_2$

In antiquity, several geometric problems concerned the (in)feasibility of constructing certain numbers with compass and straight edge. For example it was unknown to the Greeks that it is in general impossible to trisect a given angle. Using the field notion and field theory allows these problems to be settled. To do so, the field of constructible numbers is considered. It contains, on the plane, the points 0 and 1, and all complex numbers that can be constructed from these two by a finite number of construction steps using only compass and straight edge. This set, endowed with the usual addition and multiplication of complex numbers does form a field. For example, multiplying two (real) numbers r_1 and r_2 that have already been constructed can be done using construction at the right, based on the intercept theorem. This way, the obtained field F contains all rational numbers, but is bigger than Q, because for any $f \in F$, the square root of f is also a constructible number.

Finite Fields

Finite fields (also called *Galois fields*) are fields with finitely many elements. The above introductory example F_4 is a field with four elements. Highlighted in the multiplication and addition tables above is the field F_2 consisting of two elements 0 and 1. This is the smallest field, because by definition a field has at least two distinct elements $1 \neq 0$. Interpreting the addition and multiplication in this latter field as XOR and AND operations, this field finds applications in computer science, especially incryptography and coding theory.

In a finite field there is necessarily an integer n such that $1 + 1 + \cdots + 1$ (n repeated terms) equals 0. It can be shown that the smallest such n must be a prime number, called the *characteristic* of the field. If a (necessarily infinite) field has the property that $1 + 1 + \cdots + 1$ is never zero, for any number of summands, such as in Q, for example, the characteristic is said to be zero.

A basic class of finite fields are the fields F_p with p elements (p a prime number):

$$F_p = Z/pZ = \{0,\ 1,\ \ldots,\ p - 1\},$$

where the operations are defined by performing the operation in the set of integers Z, dividing by p and taking the remainder. A field K of characteristic p necessarily contains F_p,and therefore may be viewed as a vector space over F_p, of finite dimension if K is finite. Thus a finite field K has prime power order, i.e., K has $q = p^n$ elements (where $n > 0$ is the number of elements in a basis of K over F_p). By developing more field theory, in particular the notion of the splitting field of a polynomial f over a field K, which is the smallest field containing K and all roots of f, one can show that two finite fields with the same number of elements are isomorphic, i.e., there is a one-to-one mapping of one field onto the other that preserves multiplication and addition. Thus we may speak of *the* finite field with q elements, usually denoted by F_q or GF(q).

Field of Functions

Given a geometric object X, one can consider functions on such objects. Adding and multiplying them pointwise, i.e., $(f \cdot g)(x) = f(x) \cdot g(x)$ this leads to a field. However, due to the presence of possible zeros, i.e., points $x \in X$ where $f(x) = 0$, one has to take poles into account, i.e., formally allowing $f(x) = \infty$.

If X is an algebraic variety over F, then the rational functions $X \to F$, i.e., functions defined almost everywhere, form a field, the function field of X. Likewise, if X is a Riemann surface, then themeromorphic functions $S \to$ C form a field. Under certain circumstances, namely when S is compact, S can be reconstructed from this field.

Local and Global Fields

Another important distinction in the realm of fields, especially with regard to number theory, are local fields and global fields. Local fields are completions of global fields at a given place. For example, Q is a global field, and the attached local fields are Q_p and R (Ostrowski's theorem). Algebraic number fields and function fields over F_q are further global fields. Studying arithmetic questions in global fields may sometimes be done by looking at the corresponding questions locally—this technique is calledlocal-global principle.

Some First Theorems

- Every finite subgroup of the multiplicative group $F^{\times}$ is cyclic. This applies in particular to $F_q^{\times}$, it is cyclic of order q - 1. In the introductory example, a generator of $F_4^{\times}$ is the element A.
- From the point of view of algebraic geometry, fields are points, because the spectrum *Spec F* has only one point, corresponding to the 0-ideal. This entails that a commutative ring is a field if and only if it has no ideals except {0} and itself. Equivalently, an integral domain is field if and only if its Krull dimension is 0.
- Isomorphism extension theorem.

Constructing Fields

Closure Operations: Assuming the axiom of choice, for every field F, there exists a field F, called the algebraic closure of$F$, which contains F, is algebraic over F, which means that any element x of F satisfies a polynomial equation

$f_n x^n + f_{n-1} x^{n-1} + \cdots + f_1 x + f_0 = 0$, with coefficients $f_n, \ldots, f_0 \in F$, and is algebraically closed, i.e., any such polynomial does have at least one solution in *F*. The algebraic closure is unique up to isomorphism inducing the identity on *F*. However, in many circumstances in mathematics, it is not appropriate to treat *F* as being uniquely determined by *F*, since the isomorphism above is not itself unique. In these cases, one refers to such a *F* as *an* algebraic closure of *F*. A similar concept is the separable closure, containing all roots of separable polynomials, instead of all polynomials.

For example, if *F* = Q, the algebraic closure Q is also called *field of algebraic numbers*. The field of algebraic numbers is an example of an algebraically closed field of characteristic zero; as such it satisfies the same first-order sentences as the field of complex numbers C.

In general, all algebraic closures of a field are isomorphic. However, there is in general no preferable isomorphism between two closures. Likewise for separable closures.

Subfields and Field Extensions

A *subfield* is, informally, a small field contained in a bigger one. Formally, a subfield *E* of a field *F* is asubset containing 0 and 1, closed under the operations +, -, · and multiplicative inverses and with its own operations defined by restriction. For example, the real numbers contain several interesting subfields: the real algebraic numbers, the computable numbers and the rational numbers are examples.

The notion of field extension lies at the heart of field theory, and is crucial to many other algebraic domains. A field extension *F* / *E* is simply a field *F* and a subfield *E* ,⊂ *F*. Constructing such a field extension *F* / *E* can be done by "adding new elements" or *adjoining elements* to the field *E*. For example, given a field *E*, the set *F* = *E*(*X*) of rational functions, i.e., equivalence classes of expressions of the kind

$$\frac{p(X)}{q(X)},$$

where *p*(*X*) and *q*(*X*) are polynomials with coefficients in *E*, and *q* is not the zero polynomial, forms a field. This is the simplest example of a transcendental extension of *E*. It also is an example of adomain (the ring of polynomials $_E$ in this case) being embedded into its field of fractions E(X).

The ring of formal power series E[[X]] is also a domain, and again the (equivalence classes of) fractions of the form *p*(*X*)/ *q*(*X*) where p and q are elements of E[[X]] form the field of fractions for E[[X]]. This field is actually the ring of Laurent series over the field E, denoted E[[X]].

In the above two cases, the added symbol X and its powers did not interact with elements of E. It is possible however that the adjoined symbol may interact with E.

This idea will be illustrated by adjoining an element to the field of real numbers R. As explained above, C is an extension of R. C can be obtained from R by adjoining the imaginary symbol i which satisfies i^2 = "1. The result is that R[i]=C. This is different from adjoining the symbol X to R, because in that case, the powers of X are all distinct objects, but here, i^2=-1 is actually an element of R.

Another way to view this last example is to note that i is a zero of the polynomial $p(X) = X^2 + 1$. The quotient ring $R[X]/(X^2+1)$ can be mapped onto C using the map $\overline{a+bX} \rightarrow a+ib$. Since the ideal($X^2$+1) is generated by a polynomial irreducible over R, the ideal is maximal, hence the quotient ring is a field. This nonzero ring map from the quotient to C is necessarily an isomorphism of rings.

The above construction generalises to any irreducible polynomial in the polynomial ring $E[X]$, i.e., a polynomial $p(X)$ that cannot be written as a product of non-constant polynomials. The quotient ring $F = E[X] / (p(X))$, is again a field.

Alternatively, constructing such field extensions can also be done, if a bigger container is already given. Suppose given a field E, and a field G containing E as a subfield, for example G could be the algebraic closure of E. Let x be an element of G not in E.

Then there is a smallest subfield of G containing E and x, denoted $F = E(x)$ and called *field extension F / E generated by x in G*. Such extensions are also called *simple extensions*. Many extensions are of this type. For instance, Q(i) is the subfield of C consisting of all numbers of the form $a + bi$ where both a and b are rational numbers.

One distinguishes between extensions having various qualities. For example, an extension K of a field k is called *algebraic*, if every element of K is a root of some polynomial with coefficients in k. Otherwise, the extension is called *transcendental*. The aim of Galois theory is the study of *algebraic extensions* of a field.

Rings vs Fields

Adding multiplicative inverses to an integral domain R yields the field of fractions of R. For example, the field of fractions of the integers Z is just Q. Also, the field $F(X)$ is the quotient field of the ring of polynomials $F[X]$. "Getting back" the ring from the field is sometimes possible.

Another method to obtain a field from a commutative ring R is taking the quotient R / m, where m is any maximal ideal of R. The above construction of $F = E[X] / (p(X))$, is an example, because the irreducibility of the polynomial $p(X)$ is equivalent to the maximality of the ideal generated by this polynomial. Another example are the finite fields $F_p = Z / pZ$.

Ultraproducts

If I is an index set, U is an ultrafilter on I, and F_i is a field for every i in I, the ultraproduct of the F_i with respect to U is a field.

For example, a non-principal ultraproduct of finite fields is a pseudo finite field; i.e., a PAC field having exactly one extension of any degree. This construction is important to the study of the elementary theory of finite fields.

Galois Theory

Galois theory aims to study the algebraic extensions of a field by studying the symmetry in the arithmetic operations of addition and multiplication. The fundamental theorem of Galois theory shows that there is a strong relation between the structure of the symmetry group and the set of algebraic extensions.

In the case where F / E is a finite (Galois) extension, Galois theory studies the algebraic extensions of E that are subfields of F. Such fields are called intermediate extensions. Specifically, the Galois group of F over E, denoted Gal(F/E), is the group of field automorphisms of F that are trivial on E (i.e., the bijections $\sigma : F \to F$ that preserve addition and multiplication and that send elements of E to themselves), and the fundamental theorem of Galois theory states that there is a one-to-one correspondence between subgroups of Gal(F/E) and the set of intermediate extensions of the extension F/E. The theorem, in fact, gives an explicit correspondence and further properties.

To study all (separable) algebraic extensions of E at once, one must consider the absolute Galois group of E, defined as the Galois group of the separable closure, E^{sep}, of E over E (i.e., Gal(E^{sep}/E). It is possible that the degree of this extension is infinite (as in the case of E = Q). It is thus necessary to have a notion of Galois group for an infinite algebraic extension. The Galois group in this case is obtained as a "limit" (specifically an inverse limit) of the Galois groups of the finite Galois extensions ofE. In this way, it acquires a topology. The fundamental theorem of Galois theory can be generalised to the case of infinite Galois extensions by taking into consideration the topology of the Galois group, and in the case of E^{sep}/E it states that there this a one-to-one correspondence between closed subgroups of Gal(E^{sep}/E) and the set of all separable algebraic extensions of E (technically, one only obtains those separable algebraic extensions of E that occur as subfields of the chosen separable closure E^{sep}, but since all separable closures of E are isomorphic, choosing a different separable closure would give the same Galois group and thus an "equivalent" set of algebraic extensions).

Generalisations

There are also proper classes with field structure, which are sometimes called Fields, with a capital F:

- The surreal numbers form a Field containing the reals, and would be a field except for the fact that they are a proper class, not a set. The set of all surreal numbers with birthday smaller than some inaccessible cardinal form a field.
- The nimbers form a Field. The set of nimbers with birthday smaller than 2^{2n}, the nimbers with birthday smaller than any infinite cardinal are all examples of fields.

In a different direction, differential fields are fields equipped with a derivation. For example, the field R(X), together with the standard derivative of polynomials forms a differential field. These fields are central to differential Galois theory. Exponential fields, meanwhile, are fields equipped with an exponential function that provides a homomorphism between the additive and multiplicative groups within the field. The usual exponential function makes the real and complex numbers exponential fields, denoted R_{exp} and C_{exp} respectively.

Generalising in a more categorical direction yields the field with one element and related objects.

Exponentiation

One does not in general study generalisations of field with *three* binary operations. The familiar addition/subtraction, multiplication/division, exponentiation/root-extraction operations from the natural numbers to the reals, each built up in terms of iteration of the last, mean that generalising exponentiation as a binary operation is tempting, but has generally not proven fruitful; instead, an exponential field assumes a unary exponential function from the additive group to the multiplicative group, not a partially defined binary function. Note that the exponential operation of a^b is neither associative nor commutative, nor has a unique inverse (±2 are both square roots of 4, for instance), unlike addition and multiplication, and further is not defined for many pairs—for example, $(-1)^{1/2} = \sqrt{-1}$ does not define a single number. These all show that even for rational numbers exponentiation is not nearly as well-behaved as addition and multiplication, which is why one does not in general axiomatise exponentiation.

Applications

The concept of a field is of use, for example, in defining vectors and matrices, two structures in linear algebra whose components can be elements of an arbitrary field. Finite fields are used in number theory, Galois theory, coding theory and combinatorics; and again the notion of algebraic extension is an important tool. Fields of characteristic 2 are useful in computer science.

Chapter 7

Functional Analysis

Functional analysis is a branch of mathematical analysis, the core of which is formed by the study of vector spaces endowed with some kind of limit-related structure (e.g. inner product, norm, topology, etc.) and the linear operators acting upon these spaces and respecting these structures in a suitable sense. The historical roots of functional analysis lie in the study of spaces of functions and the formulation of properties of transformations of functions such as the Fourier transform as transformations defining continuous, unitary etc. operators between function spaces. This point of view turned out to be particularly useful for the study of differential and integral equations.

The usage of the word *functional* goes back to the calculus of variations, implying a function whose argument is a function and the name was first used in Hadamard's 1910 book on that subject. However, the general concept of functional had previously been introduced in 1887 by the Italian mathematician and physicist Vito Volterra. The theory of nonlinear functionals was continued by students of Hadamard, in particular Fréchet and Lévy. Hadamard also founded the modern school of linear functional analysis further developed by Riesz and the group of Polish mathematicians around Stefan Banach.

In modern introductory texts to functional analysis, the subject is seen as the study of vector spaces endowed with a topology, in particular infinite dimensional spaces. In contrast, linear algebra deals mostly with finite dimensional spaces, or does not use topology. An important part of functional analysis is the extension of the theory of measure, integration, and probability to infinite dimensional spaces, also known as infinite dimensional analysis.

Normed Vector Spaces

The basic and historically first class of spaces studied in functional analysis are complete normed vector spaces over the real or complex numbers. Such spaces are called Banach spaces. An important example is a Hilbert space, where the norm arises from an inner product. These spaces are of fundamental importance in many areas, including the mathematical formulation of quantum mechanics.

More generally, functional analysis includes the study of Fréchet spaces and other topological vector spaces not endowed with a norm.

An important object of study in functional analysis are the continuous linear operators defined on Banach and Hilbert spaces. These lead naturally to the definition of C*-algebras and other operator algebras.

Hilbert Spaces

Hilbert spaces can be completely classified: there is a unique Hilbert space up to isomorphism for every cardinality of the orthonourmal basis. Finite-dimensional Hilbert spaces are fully understood inlinear algebra, and infinite-dimensional separable Hilbert spaces are isomorphic to $\ell^2(\aleph_0)$.

Separability being important for applications, functional analysis of Hilbert spaces consequently mostly deals with this space. One of the open problems in functional analysis is to prove that every bounded linear operator on a Hilbert space has a proper invariant subspace. Many special cases of this invariant subspace problem have already been proven.

Banach Spaces

General Banach spaces are more complicated than Hilbert spaces, and cannot be classified in such a simple manner as those. In particular, Banach spaces lack a notion analogous to an orthonourmal basis.

Examples of Banach spaces are L^p-spaces for any real number $p \geq 1$. Given also a measure μon set *X*, then $L^p(X)$, sometimes also denoted $L^p(X,\mu)$ or $L^p(\mu)$, has as its vectors equivalence classes [f] of measurable functions whose absolute value's *p*-th power has finite integral, that is, functions f for which one has

$$\int_X |f(x)|^p \, d\mu(x) < +\infty .$$

If μ is the counting measure, then the integral may be replaced by a sum. That is, we require

$$\sum_{x \in X} |f(x)|^p < +\infty .$$

Then it is not necessary to deal with equivalence classes, and the space is denoted $\ell^p(X)$, written more simply ℓ^p in the case when X is the set of non-negative integers.

In Banach spaces, a large part of the study involves the dual space: the space of all continuous linear maps from the space into its underlying field, so-called functionals.

A Banach space can be canonically identified with a subspace of its bidual, which is the dual of its dual space. The corresponding map is an isometry but in general not onto. A general Banach space and its bidual need not even be isometrically isomorphic in any way, contrary to the finite-dimensional situation. This is explained in the dual space article.

Also, the notion of derivative can be extended to arbitrary functions between Banach spaces.

Major and Foundational Results

Important results of functional analysis include:

- The uniform boundedness principle (also known as Banach–Steinhaus theorem) applies to sets of operators with uniform bounds.
- One of the spectral theorems (there are indeed more than one) gives an integral formula for the normal operators on a Hilbert space. This theorem is of central importance for the mathematical formulation of quantum mechanics.
- The Hahn–Banach theorem extends functionals from a subspace to the full space, in a norm-preserving fashion. An implication is the non-triviality of dual spaces.
- The open mapping theorem and closed graph theorem.

Foundations of Mathematics Considerations

Most spaces considered in functional analysis have infinite dimension. To show the existence of a vector space basis for such spaces may require Zorn's lemma. However, a somewhat different concept, Schauder basis, is usually more relevant in functional analysis. Many very important theorems require the Hahn–Banach theorem, usually proved using axiom of choice, although the strictly weaker Boolean prime ideal theorem suffices. The Baire category theorem, needed to prove many important theorems, also requires a form of axiom of choice.

Points of View

Functional analysis in its present form includes the following tendencies:

- *Abstract analysis*. An approach to analysis based on topological groups, topological rings, and topological vector spaces;
- *Geometry of Banach spaces* contains many topics. One is combinatorial approach connected with Jean Bourgain; another is a characterisation of Banach spaces in which various forms of the law of large numbers hold.
- *Noncommutative geometry*. Developed by Alain Connes, partly building on earlier notions, such as George Mackey's approach to ergodic theory.
- *Connection with quantum mechanics*. Either narrowly defined as in mathematical physics, or broadly interpreted by, e.g. Israel Gelfand, to include most types of representation theory.

In mathematics, with 2- or 3-dimensional vectors with real-valued entries, the idea of the "length" of a vector is intuitive and can easily be extended to any real vector space R^n. The following properties of "vector length" are crucial.

1. The zero vector, 0, has zero length; every other vector has a positive length.

   $||x||>0$ if $x \neq 0$
2. Multiplying a vector by a positive number changes its length without changing its direction. Moreover,

   $||\alpha x||=|\alpha|\,||x||$ for any scalar α.
3. The triangle inequality holds. That is, taking norms as distances, the distance from point A through B to C is never shorter than going directly from A to C, or the shortest distance between any two points is a straight line.

$||x+y||\leq||x||+||y||$ for any vectors x and y. (triangle inequality)

The generalisation of these three properties to more abstract vector spaces leads to the notion of norm. A vector space on which a norm is defined is then called a normed vector space. Normed vector spaces are central to the study of linear algebra and functional analysis.

A seminormed vector space is a pair (V,p) where V is a vector space and p a seminorm on V. A normed vector space is a pair $(V, \|\cdot\|)$ where V is a vector space and $\|\cdot\|$ a norm on V.

We often omit p or $\|\cdot\|$ and just write V for a space if it is clear from the context what (semi) norm we are using.

In a more general sense, a vector norm can be taken to be any real-valued function that satisfies these three properties. The properties 1. and 2. together imply that

$$||x|| = 0 \text{ if and only if } x = 0.$$

A useful variation of the triangle inequality is

$$||x-y|| \geq |\,||x|| - ||y||\,| \text{ for any vectors x and y.}$$

This also shows that a vector norm is a continuous function.

Topological Structure

If $(V, \|\cdot\|)$ is a normed vector space, the norm $\|\cdot\|$ induces a metric (a notion of *distance*) and therefore a topology on V. This metric is defined in the natural way: the distance between two vectors u and v is given by u-v . This topology is precisely the weakest topology which makes · continuous and which is compatible with the linear structure of V in the following sense:

1. The vector addition $+ : V \times V \to V$ is jointly continuous with respect to this topology. This follows directly from the triangle inequality.
2. The scalar multiplication $\|\cdot\| : K \times V \to V$, where K is the underlying scalar field of V, is jointly continuous. This follows from the triangle inequality and homogeneity of the norm.

Similarly, for any semi-normed vector space we can define the distance between two vectors u and v as u-v. This turns the seminormed space into a pseudometric space (notice this is weaker than a metric) and allows the definition of notions such as continuity and convergence. To put it more abstractly every semi-normed vector space is a topological vector space and thus carries a topological structure which is induced by the semi-norm.

Of special interest are complete normed spaces called Banach spaces. Every normed vector space Vsits as a dense subspace inside a Banach space; this Banach space is essentially uniquely defined by V and is called the *completion* of V.

All norms on a finite-dimensional vector space are equivalent from a topological viewpoint as they induce the same topology (although the resulting metric spaces need not be the same). And since any Euclidean space is complete, we can thus conclude that all finite-dimensional normed vector spaces are Banach spaces.

A normed vector space V is locally compact if and only if the unit ball $B = \{x : \ x \ \leq 1\}$ is compact, which is the case if and only if V is finite-dimensional; this is a consequence of Riesz's lemma. (In fact, a more general result is true: a topological vector space is locally compact if and only if it is finite-dimensional. The point here is that we don't assume the topology comes from a norm.)

The topology of a seminormed vector has many nice properties. Given a neighbourhood system $\mathcal{N}(0)$ around 0 we can construct all other neighbourhood systems as

$$\mathcal{N}(\mathrm{x}) = \mathrm{x} + \mathcal{N}(0) := \{\mathrm{x} + \mathrm{N} \mid \mathrm{N} \in \mathcal{N}(0)\}$$

with

$$\mathrm{x} + \mathrm{N} := \{\mathrm{x} + \mathrm{n} \mid \mathrm{n} \in \mathrm{N}\}.$$

Moreover there exists a neighbourhood basis for 0 consisting of absorbing and convex sets. As this property is very useful in functional analysis, generalisations of normed vector spaces with this property are studied under the name locally convex spaces.

Linear Maps and Dual Spaces

The most important maps between two normed vector spaces are the continuous linear maps. Together with these maps, normed vector spaces form a category.

The norm is a continuous function on its vector space. All linear maps between finite dimensional vector spaces are also continuous.

An isometry between two normed vector spaces is a linear map f which preserves the norm (meaning f(v) = v for all vectors v). Isometries are always continuous and injective. A surjective isometry between the normed vector spaces V and W is called an isometric isomorphism, and V and W are called isometrically isomorphic. Isometrically isomorphic normed vector spaces are identical for all practical purposes.

When speaking of normed vector spaces, we augment the notion of dual space to take the norm into account. The dual V ' of a normed vector space V is the space of all continuous linear maps from V to the base field (the complexes or the reals) — such linear maps are called "functionals". The norm of a functional φ is defined as the supremum of $|\varphi(v)|$ where v ranges over all unit vectors (i.e. vectors of norm 1) in V. This turns V ' into a normed vector space. An important theorem about continuous linear functionals on normed vector spaces is the Hahn–Banach theorem.

Normed Spaces as Quotient Spaces of Seminormed Spaces

The definition of many normed spaces (in particular, Banach spaces) involves a seminorm defined on a vector space and then the normed space is defined as the quotient space by the subspace of elements of seminorm zero. For instance, with the L^p spaces, the function defined by

$$\|f\|_p = \left(\int |f(x)|^p\, dx\right)^{1/p}$$

is a seminorm on the vector space of all functions on which the Lebesgue integral on the right hand side is defined and finite. However, the seminorm is equal to zero for any function supported on a set of Lebesgue measure zero. These functions form a subspace which we "quotient out", making them equivalent to the zero function.

Finite Product Spaces

Given *n* seminormed spaces X_i with seminorms q_i we can define the product space as

$$X := \prod_{i=1}^{n} X_i$$

with vector addition defined as

$$(x_1,\ldots,x_n)+(y_1,\ldots,y_n) := (x_1+y_1,\ldots,x_n+y_n)$$

and scalar multiplication defined as

$$\alpha(x_1,\ldots,x_n) := (\alpha x_1,\ldots,\alpha x_n).$$

We define a new function *q*

$$q : X \mapsto \mathbb{R}$$

for example as

$$q : (x_1,\ldots,x_n) \to \sum_{i=1}^{n} q_i(x_i).$$

which is a seminorm on *X*. The function *q* is a norm if and only if all q_i are norms.

More generally, for each real $p \geq 1$ we have the seminorm:

$$q:(x_1,\ldots,x_n)\to\left(\sum_{i=1}^{n}q_i(x_i)^p\right)^{\frac{1}{p}}.$$

For each p this defines the same topological space.

A straightforward argument involving elementary linear algebra shows that the only finite-dimensional seminormed spaces are those arising as the product space of a normed space and a space with trivial seminorm. Consequently, many of the more interesting examples and applications of seminormed spaces occur for infinite-dimensional vector spaces.

Banach Space

In mathematics, Banach spaces (pronounced [Èbanax]) is the name for complete normed vector spaces, one of the central objects of study in functional analysis. A complete normed vector space is a vector space V with a norm $||\cdot||$ such that every Cauchy sequence (with respect to the metric $d(x, y) = ||x - y||$) in V has a limit in V (with respect to the topology induced by that metric). As for general vector spaces, a Banach space over the real numbers is called a real Banach space, and a Banach space over the complex numbers is called a complex Banach space.

Many of the infinite-dimensional function spaces studied in analysis are Banach spaces, including spaces of continuous functions (continuous functions on a compact Hausdorff space), spaces of Lebesgue integrable functions known as L^p spaces, and spaces of holomorphic functions known asHardy spaces. They are the most commonly used topological vector spaces, and their topology comes from a norm.

They are named after the Polish mathematician Stefan Banach, who introduced them in 1920–1922 along with Hans Hahn and Eduard Helly.

Examples

Throughout, let K stand for one of the fields R or C.

The familiar Euclidean spaces K^n, *where the Euclidean norm of* $x = (x_1, \ldots, x_n)$ *is given by* $||x|| = (x_1^2+\ldots+x_n^2)^{1/2}$, *are Banach spaces. Hence every finite-dimensional K vector space becomes a Banach space being endowed with an arbitrary norm, since all norms are equivalent on a finite-dimensional Kvector space.*

Consider the space of all continuous functions $f : [a, b] \to K$ defined on a closed interval [a, b]. This space becomes a Banach space if an appropriate norm $||f||$, is defined in it. Such a norm may be defined as $||f|| = \sup \{ |f(x)| : x$ " $[a, b] \}$, known as the supremum norm. This is indeed a well-defined norm, since continuous functions defined on a closed interval are bounded.

Since f is a continuous function on a closed interval, then it is bounded and the supremum in the above definition is attained by the Weierstrass extreme value theorem, so we can replace the supremum by the maximum. In this case, the norm is also called the maximum norm.

The space is complete under this norm, and the resulting Banach space is denoted by C[a, b]. This example can be generalised to the space C(X) of all continuous functions $X \to K$, where X is a compact space, or to the space of all bounded continuous functions $X \to K$, where X is anytopological space, or indeed to the space B(X) of all bounded functions $X \to K$, where X is any set. In all these examples, we can even multiply functions and stay in the same space: all these examples are in fact unital Banach algebras.

For any open set $\Omega \subseteq C$, the set $A(\Omega)$ of all bounded, analytic functions $u : \Omega \to C$ is a complex Banach space with respect to the supremum norm. The fact that uniform limits of analytic functions are analytic is an easy consequence of Morera's theorem.

If $p \geq 0$ is a real number, we can consider the space of all infinite sequences $(x_1, x_2, x_3, \ldots)$ of elements in K such that the infinite series $\varepsilon_i \, |x_i|^p$ is finite. The p-th root of this series' value is then defined to be the p-norm of the sequence. The space, together with this norm, is a Banach space; it is denoted by ℓ^p.

The Banach space ℓ^∞ consists of all bounded sequences of elements in K; the norm of such a sequence is defined to be the supremum of the absolute values of the sequence's members.

Again, if $p \geq 1$ is a real number, we can consider all functions $f : [a, b] \to K$ such that $|f|^p$ is Lebesgue integrable. The p-th root of this integral is then defined to be the norm of f. By itself, this space is not a Banach space because there are non-zero functions whose norm is zero. We define an equivalence relation as follows: f and g are equivalent if and only if the norm of f-g is zero. The set of equivalence classes then forms a Banach space; it is denoted by $L^p([a, b])$. It is crucial to use the Lebesgue integral and not the Riemann integral here, *because the Riemann integral would not yield a complete space. These examples can be generalised.*

If X *and* Y *are two Banach spaces, then we can form their direct sum* X ⊕ Y, *which has a natural topological vector space structure but no canonical norm. However, it is again a Banach space for several equivalent norms, for example*

$$||x \oplus y|| = \left(||x||^p + ||y||^p \right)^{1/p}, \quad 1 \le p \le \infty.$$

This construction can be generalised to define ℓ^p-direct sums of arbitrarily many Banach spaces. When there is an infinite number of non-zero summands, the space obtained in this way depends upon *p*.

If *M* is a closed linear subspace of the Banach space *X*, then the quotient space *X* / *M* is again a Banach space.

Every inner product gives rise to an associated norm. The inner product space is called a Hilbert space if its associated norm is complete. Thus every Hilbert space is a Banach space by definition. The converse statement also holds under certain conditions.

Linear Operators

If V *and* W *are Banach spaces over the same ground field K, the set of all continuous K-linear maps* $A : V \to W$ *is denoted by L(*V, W*). In infinite-dimensional spaces, not all linear maps are automatically continuous. In general, a linear mapping on a normed space is continuous if and only if it is boundedon the closed unit ball. Thus the vector space L(*V, W*) can be given the operator norm*

$$||A|| = \sup\{||Ax||_W \mid x \in V, ||x||_V \le 1\}.$$

With respect to this norm, L(*V*, *W*) is a Banach space. This is also true under the less restrictive condition that *V* be a normed space.

When *V* = *W*, the space L(*V*) = L(*V*, *V*) forms a unital Banach algebra; the multiplication operation is given by the composition of linear maps.

Dual Space

If V *is a Banach space and K is the underlying field (either the real or the complex numbers), then K is itself a Banach space (using the absolute value as norm) and we can define the* dual space V2 *as* V2 = *L(*V, *K), the space of continuous linear maps into K. This is again a Banach space (with the operator norm). It can be used to define a new topology on* V*: the weak topology.*

Note that the requirement that the maps be continuous is essential; if V *is infinite-dimensional, there exist linear maps which are not continuous, and therefore not bounded, so the space* V^* *of linear maps into K is not a Banach space. The space* V^* *(which may be called the* algebraic dual space *to distinguish it from* V'*) also induces a weak topology which is finer than that induced by the continuous dual since* $V' \subseteq V^*$.

There is a natural map F *from* V *to* V2 2 *(the dual of the dual) defined by*

$$F(x)(f) = f(x)$$

for all x in V and f in V'. Because $F(x)$ is a map from V' to K, it is an element of V''. The map $F: x \to F(x)$ is thus a map $V \to V''$. As a consequence of the Hahn–Banach theorem, this map is injective, and isometric; if it is also surjective, then the Banach space V is called reflexive. Reflexive spaces have many important geometric properties. A space is reflexive if and only if its dual is reflexive, which is the case if and only if its unit ball is compact in the weak topology.

For example, ℓ^p is reflexive for $1 < p < \infty$ but ℓ^1 and ℓ^∞ are not reflexive. When $p < \infty$, the dual of ℓ^p is ℓ^q where p and q are related by the formula $1/p + 1/q = 1$.

Relationship to Hilbert Spaces

As mentioned above, every Hilbert space is a Banach space because, by definition, a Hilbert space is complete with respect to the norm associated with its inner product, where a norm and an inner product are said to be associated if $||v||^2 = (v,v)$ for all v.

The converse is not always true; not every Banach space is a Hilbert space. A necessary and sufficient condition for a Banach space V *to be associated to an inner product (which will then necessarily make* V *into a Hilbert space) is the parallelogram identity:*

$$||u+v||^2 + ||u-v||^2 = 2(||u||^2 + ||v||^2)$$

for all u and v in V, and where $||*||$ is the norm on V. So, for example, while R^n is a Banach space with respect to *any* norm defined on it, it is only a Hilbert space with respect to the Euclidean norm. Similarly, as an infinite-dimensional example, the Lebesgue space L^p is always a Banach space but is only a Hilbert space when $p = 2$.

If the norm of a Banach space satisfies this identity, the associated inner product which makes it into a Hilbert space is given by the polarisation identity. If V is a real Banach space, then the polarisation identity is

$$\langle u, v\rangle = \frac{1}{4}(||u+v||^2 - ||u-v||^2)$$

whereas if V is a complex Banach space, then the polarisation identity is given by (assuming that scalar product is linear in first argument):

$$\langle u, v\rangle = \frac{1}{4}\left(||u+v||^2 - ||u-v||^2 + i(||u+iv||^2 - ||u-iv||^2)\right).$$

The necessity of this condition follows easily from the properties of an inner product. It is sufficient—that the parallelogram law implies that the form defined by the polarisation identity is indeed a complete inner product—one verifies algebraically that this form is additive, whence it follows by induction that the form is linear over the integers and rationals. Then since every real is the limit of some Cauchy sequence of rationals, the completeness of the norm extends the linearity to the whole real line. In the complex case, one can check also that the bilinear form is linear over i in one argument, and conjugate linear in the other.

Hamel Dimension

It follows from the completeness of Banach spaces and the Baire category theorem that a Hamel basis of an infinite-dimensional Banach space is uncountable.

Derivatives

Several concepts of a derivative may be defined on a Banach space.

Generalisations

Several important spaces in functional analysis, for instance the space of all infinitely often differentiable functions $R \to R$ or the space of all distributions on R, are complete but are not normed vector spaces and hence not Banach spaces. In Fréchet spaces one still has a complete metric, while LF-spaces are complete uniform vector spaces arising as limits of Fréchet spaces.

Hilbert Space

The mathematical concept of a Hilbert space, named after David Hilbert, generalises the notion of Euclidean space. It extends the methods of vector algebra and calculus from the two-dimensional Euclidean plane and three-dimensional space to spaces with any finite or infinite number of dimensions. A Hilbert space is an abstract vector space possessing the structure of an inner product that allows length and angle to be measured. Furthermore, Hilbert spaces are required to be complete, a property that stipulates the existence of enough limits in the space to allow the techniques of calculus to be used.

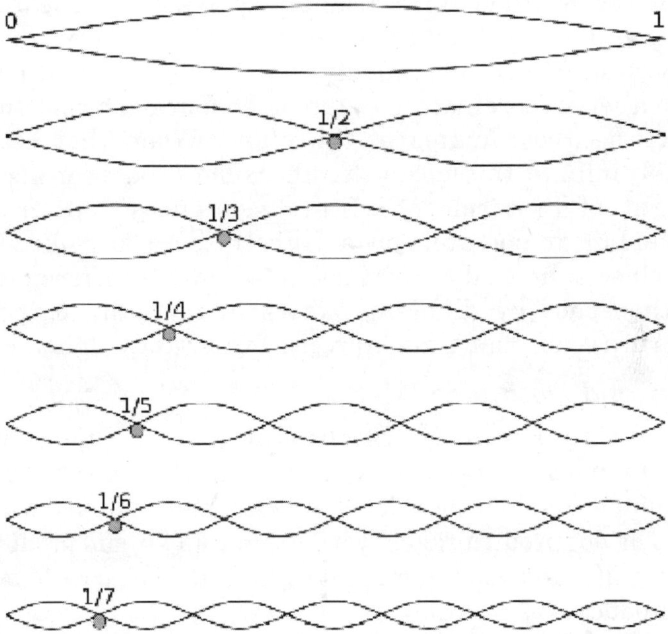

Figure 1: Hilbert spaces can be used to study the harmonics of vibrating strings.

Hilbert spaces arise naturally and frequently in mathematics, physics, and engineering, typically as infinite-dimensional function spaces. The earliest Hilbert spaces were studied from this point of view in the first decade of the 20th century by David Hilbert, Erhard Schmidt, and Frigyes Riesz.

They are indispensable tools in the theories of partial differential equations, quantum mechanics, Fourier analysis (which includes applications to signal processing and heat transfer) and ergodic theory which forms the mathematical underpinning of the study of thermodynamics. John von Neumann coined the term "Hilbert space" for the abstract concept underlying many of these diverse applications.

The success of Hilbert space methods ushered in a very fruitful era for functional analysis. Apart from the classical Euclidean spaces, examples of Hilbert spaces include spaces of square-integrable functions, spaces of sequences, Sobolev spaces consisting of generalised functions, and Hardy spaces of holomorphic functions.

Geometric intuition plays an important role in many aspects of Hilbert space theory. Exact analogues of the Pythagorean theorem and parallelogram law hold in a Hilbert space. At a deeper level, perpendicular projection onto a subspace (the analogue of "dropping the altitude" of a triangle) plays a significant role in optimisation problems and other aspects of the theory. An element of a Hilbert space can be uniquely specified by its coordinates with respect to a set of coordinate axes (an orthonourmal basis), in analogy with Cartesian coordinates in the plane. When that set of axes is countably infinite, this means that the Hilbert space can also usefully be thought of in terms of infinite sequences that are square-summable. Linear operators on a Hilbert space are likewise fairly concrete objects: in good cases, they are simply transformations that stretch the space by different factors in mutually perpendicular directions in a sense that is made precise by the study of their spectrum.

Definition and Illustration

Motivating Example: Euclidean Space: One of the most familiar examples of a Hilbert space is the Euclidean space consisting of three-dimensional vectors, denoted by R^3, and equipped with the dot product. The dot product takes two vectors x and y, and produces a real number x·y. If x and y are represented in Cartesian coordinates, then the dot product is defined by

$$(x_1, x_2, x_3) \cdot (y_1, y_2, y_3) = x_1 y_1 + x_2 y_2 + x_3 y_3.$$

The dot product satisfies the properties:

1. It is symmetric in x and y: x·y = y·x.
2. It is linear in its first argument: $(ax_1 + bx_2) \cdot y = ax_1 \cdot y + bx_2 \cdot y$ for any scalars a, b, and vectorsx_1 , x_2, and y.
3. It is positive definite: for all vectors x, $x \cdot x \geq 0$ with equality if and only if x = 0.

An operation on pairs of vectors that, like the dot product, satisfies these three properties is known as a (real) inner product. A vector space equipped with such an inner product is known as a (real) inner product space. Every finite-dimensional inner product space is also a Hilbert space. The basic feature of the dot product that connects it with Euclidean geometry is that it is related to both the length (or norm) of a vector, denoted $||x||$, and to the angle θ between two vectors x and y by means of the formula

$$x \cdot y = \| x \| \| y \| \cos\theta.$$

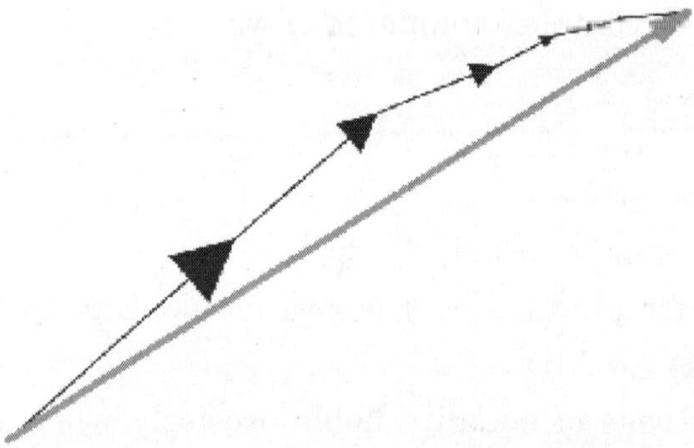

Figure 2: Completeness means that if a particle moves along the broken path (in blue) travelling a finite total distance, then the particle has a well-defined net displacement (in orange).

Multivariable calculus in Euclidean space relies on the ability to compute limits, and to have useful criteria for concluding that limits exist. A mathematical series

$$\sum_{n=0}^{\infty} x_n$$

consisting of vectors in R^3 is absolutely convergent provided that the sum of the lengths converges as an ordinary series of real numbers:

$$\sum_{k=0}^{\infty} \| x_k \| < \infty.$$

Just as with a series of scalars, a series of vectors that converges absolutely also converges to some limit vector L in the Euclidean space, in the sense that

$$\left\| L - \sum_{k=0}^{N} x_k \right\| \to 0 \quad \text{as } N \to \infty.$$

This property expresses the *completeness* of Euclidean space: that a series which converges absolutely also converges in the ordinary sense.

Definition

A Hilbert space *H* is a real or complex inner product space that is also a complete metric space with respect to the distance function induced by the inner product. To say that *H* is a complex inner product space means that *H* is a complex vector space on which there is an inner product *[x,y]* associating a complex number to each pair of elements *x,y* of *H* that satisfies the following properties:

- y,x is the complex conjugate of x,y:

  $$\langle y,x\rangle = \overline{\langle x,y\rangle}.$$

- x,y is linear in its first argument. For all complex numbers a and b,

  $$\langle ax_1 + bx_2, y\rangle = a\langle x_1, y\rangle + b\langle x_2, y\rangle.$$

- The inner product $[\cdot,\cdot]$ is positive definite:

  $$\langle x,x\rangle \geq 0$$

  where the case of equality holds precisely when $x = 0$.

It follows from properties 1 and 2 that a complex inner product is antilinear in its second argument, meaning that

$$\langle x, ay_1 + by_2\rangle = \bar{a}\langle x, y_1\rangle + \bar{b}\langle x, y_2\rangle.$$

A real inner product space is defined in the same way, except that H is a real vector space and the inner product takes real values. Such an inner product will be bilinear: that is, linear in each argument.

The norm defined by the inner product $[\cdot,\cdot]$ is the real-valued function

$$||x|| = \sqrt{\langle x,x\rangle},$$

and the distance between two points x,y in H is defined in terms of the norm by

$$d(x,y) = ||x-y|| = \sqrt{\langle x-y, x-y\rangle}.$$

That this function is a distance function means (1) that it is symmetric in x and y, (2) that the distance between x and itself is zero, and otherwise the distance between x and y must be positive, and (3) that the triangle inequality holds, meaning that the length of one leg of a triangle xyz cannot exceed the sum of the lengths of the other two legs:

$$d(x,z) \leq d(x,y) + d(y,z).$$

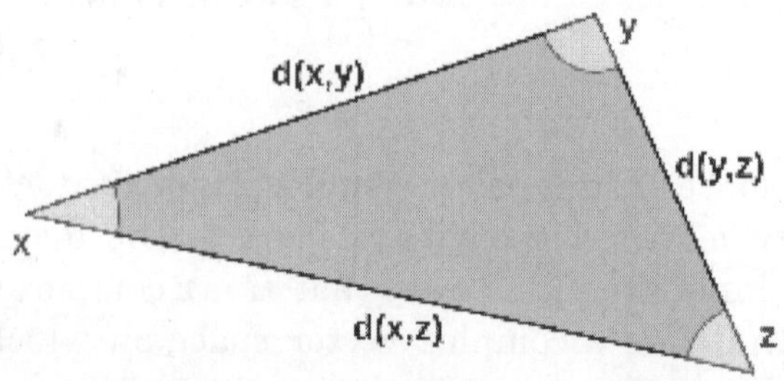

Figure 3: This last property is ultimately a consequence of the more fundamental Cauchy–Schwarz inequality, which asserts

$$|\langle x, y\rangle| \leq \|x\| \|y\|$$

with equality if and only if x and y are linearly dependent.

Relative to a distance function defined in this way, any inner product space is a metric space, and sometimes is known as a pre-Hilbert space. Any pre-Hilbert space that is additionally also acomplete space is a Hilbert space. Completeness is expressed using a form of the Cauchy criterionfor sequences in H: a pre-Hilbert space H is complete if every Cauchy sequence converges with respect to this norm to an element in the space. Completeness can be characterised by the following equivalent condition: if a series of vectors $\sum_{k=0}^{\infty} u_k$ converges absolutely in the sense that

$$\sum_{k=0}^{\infty} \|u_k\| < \infty,$$

then the series converges in H, in the sense that the partial sums converge to an element of H.

As a complete normed space, Hilbert spaces are by definition also Banach spaces. As such they are topological vector spaces, in which topological notions like the openness and closedness of subsets are well-defined. Of special importance is the notion of a closed linear subspace of a Hilbert space which, with the inner product induced by restriction, is also complete (being a closed set in a complete metric space) and therefore a Hilbert space in its own right.

Second Example: Sequence Spaces

The sequence space ℓ^2 consists of all infinite sequences $z = (z_1, z_2, \ldots)$ of complex numbers such that the series

$$\sum_{n=1}^{\infty} |z_n|^2$$

converges. The inner product on ℓ^2 is defined by

$$\langle z, w\rangle = \sum_{n=1}^{\infty} z_n \overline{w_n},$$

with the latter series converging as a consequence of the Cauchy–Schwarz inequality.

Completeness of the space holds provided that whenever a series of elements from ℓ^2 converges absolutely (in norm), then it converges to an element of ℓ^2. The proof is basic in mathematical analysis, and permits mathematical series of elements of the space to be manipulated with the same ease as series of complex numbers (or vectors in a finite-dimensional Euclidean space).

History

Prior to the development of Hilbert spaces, other generalisations of Euclidean spaces were known to mathematicians and physicists. In particular, the idea of an abstract linear space had gained some traction towards the end of the 19th century: this is a space whose elements can be added together and multiplied by scalars (such as real or complex numbers) without necessarily identifying these elements with "geometric" vectors, such as position and momentum vectors in physical systems.

Other objects studied by mathematicians at the turn of the 20th century, in particular spaces of sequences (including series) and spaces of functions, can naturally be thought of as linear spaces. Functions, for instance, can be added together or multiplied by constant scalars, and these operations obey the algebraic laws satisfied by addition and scalar multiplication of spatial vectors.

In the first decade of the 20th century, parallel developments led to the introduction of Hilbert spaces. The first of these was the observation, which arose during David Hilbert and Erhard Schmidt's study of integral equations, that two square-integrable real-valued functions f *and* g *on an interval [*a,b*] have an* inner product

$$\langle f,g\rangle = \int_a^b f(x)g(x)dx$$

which has many of the familiar properties of the Euclidean dot product. In particular, the idea of an orthogonal family of functions has meaning. Schmidt exploited the similarity of this inner product with the usual dot product to prove an analogue of the spectral decomposition for an operator of the form

$$f(x) \mapsto \int_a^b K(x,y)f(y)dy$$

where K is a continuous function symmetric in x and y. The resulting eigenfunction expansion expresses the function K as a series of the form

$$K(x,y) = \sum_n \lambda_n \varphi_n(x)\varphi_n(y)$$

where the functions φ_n are orthogonal in the sense that $[\varphi_n, \varphi_m] = 0$ for all $n \neq m$. The individual terms in this series are sometimes referred to as elementary product solutions. However, there are eigenfunction expansions which fail to converge in a suitable sense to a square-integrable function: the missing ingredient, which ensures convergence, is completeness.

The second development was the Lebesgue integral, an alternative to the Riemann integral introduced by Henri Lebesgue in 1904. The Lebesgue integral made it possible to integrate a much broader class of functions.

In 1907, Frigyes Riesz and Ernst Sigismund Fischer independently proved that the space L^2 of square Lebesgue-integrable functions is a complete metric space. As a consequence of the interplay between geometry and completeness, the 19th century results of Joseph Fourier, Friedrich Bessel and Marc-Antoine Parseval on trigonometric series easily carried over to these more general spaces, resulting in a geometrical and analytical apparatus now usually known as the Riesz–Fischer theorem.

Further basic results were proved in the early 20th century. For example, the Riesz representation theorem was independently established by Maurice Fréchet and Frigyes Riesz in 1907. John von Neumann coined the term *abstract Hilbert space* in his work on unbounded Hermitian operators.

Although other mathematicians such as Hermann Weyl and Norbert Wiener had already studied particular Hilbert spaces in great detail, often from a physically motivated point of view, von Neumann gave the first complete and axiomatic treatment of them. Von Neumann later used them in his seminal work on the foundations of quantum mechanics, and in his continued work with Eugene Wigner. The name "Hilbert space" was soon adopted by others, for example by Hermann Weyl in his book on quantum mechanics and the theory of groups.

The significance of the concept of a Hilbert space was underlined with the realisation that it offers one of the best mathematical formulations of quantum mechanics. In short, the states of a quantum mechanical system are vectors in a certain Hilbert space, the observables are hermitian operators on that space, the symmetries of the system are unitary operators, and measurements are orthogonal projections.

The relation between quantum mechanical symmetries and unitary operators provided an impetus for the development of the unitary representation theory of groups, initiated in the 1928 work of Hermann Weyl. On the other hand, in the early 1930s it became clear that certain properties of classical dynamical systems can be analysed using Hilbert space techniques in the framework of ergodic theory. The algebra of observables in quantum mechanics is naturally an algebra of operators defined on a Hilbert space, according to Werner Heisenberg's matrix mechanics formulation of quantum theory. Von Neumann began investigating operator algebras in the 1930s, as rings of operators on a Hilbert space.

The kind of algebras studied by von Neumann and his contemporaries are now known as von Neumann algebras. In the 1940s, Israel Gelfand, Mark Naimark and Irving Segal gave a definition of a kind of operator algebras called C*-algebras that on the one hand made no reference to an underlying Hilbert space, and on the other extrapolated many of the useful features of the operator algebras that had previously been studied. The spectral theorem for self-adjoint operators in particular that underlies much of the existing Hilbert space theory was generalised to C*-algebras. These techniques are now basic in abstract harmonic analysis and representation theory.

Examples

Lebesgue Spaces

Lebesgue spaces are function spaces associated to measure spaces (X, M, μ), where X is a set, M is a σ-algebra of subsets of X, and μ is a countably additive measure on M. Let $L^2(X,\mu)$ be the space of those complex-valued measurable functions on X for which the Lebesgue integral of the square of the absolute value of the function is finite, i.e., for a function f in $L^2(X,\mu)$,

$$\int_X |f|^2 \, d\mu < \infty,$$

and where functions are identified if and only if they differ only on a set of measure zero.

The inner product of functions f and g in $L^2(X,\mu)$ is then defined as

$$\langle f,g\rangle = \int_X f(t)\overline{g(t)} \, d\mu(t).$$

For f and g in L^2, this integral exists because of the Cauchy–Schwarz inequality, and defines an inner product on the space. Equipped with this inner product, L^2 is in fact complete. The Lebesgue integral is essential to ensure completeness: on domains of real numbers, for instance, not enough functions are Riemann integrable.

The Lebesgue spaces appear in many natural settings. The spaces $L^2(\mathrm{R})$ and $L^2([0,1])$ of square-integrable functions with respect to the Lebesgue measure on the real line and unit interval, respectively, are natural domains on which to define the Fourier transform and Fourier series. In other situations, the measure may be something other than the ordinary Lebesgue measure on the real line. For instance, if *w* is any positive measurable function, the space of all measurable functions *f* on the interval [0,1] satisfying

$$\int_0^1 |f(t)|^2 \, w(t)dt < \infty$$

is called the weighted L^2 space *L*2

w([0,1]), and *w* is called the weight function. The inner product is defined by

$$\langle f,g\rangle = \int_0^1 f(t)\overline{g(t)}w(t)dt.$$

The weighted space *L*2

w([0,1]) is identical with the Hilbert space $L^2([0,1],\mu)$ where the measure μ of a Lebesgue-measurable set *A* is defined by

$$\mu(A) = \int_A w(t)dt.$$

Weighted L^2 spaces like this are frequently used to study orthogonal polynomials, because different families of orthogonal polynomials are orthogonal with respect to different weighting functions.

Sobolev Spaces

Sobolev spaces, denoted by H^s or $W^{s,2}$, are Hilbert spaces. These are a special kind of function space in which differentiation may be performed, but which (unlike other Banach spaces such as the Hφlder spaces) support the structure of an inner product. Because differentiation is permitted, Sobolev spaces are a convenient setting for the theory of partial differential equations. They also form the basis of the theory of direct methods in the calculus of variations.

For *s* a non-negative integer and $\Omega, \in \mathrm{R}^n$, the Sobolev space $H^s(\Omega)$ contains L^2 functions whose weak derivatives of order up to *s* are also L^2. The inner product in $H^s(\Omega)$ is

$$\langle f,g\rangle = \int_\Omega f(x)\bar{g}(x)dx + \int_\Omega Df\cdot D\bar{g}(x)dx + \cdots + \int_\Omega D^s f(x)\cdot D^s\bar{g}(x)dx$$

where the dot indicates the dot product in the Euclidean space of partial derivatives of each order. Sobolev spaces can also be defined when *s* is not an integer.

Sobolev spaces are also studied from the point of view of spectral theory, relying more specifically on the Hilbert space structure. If Ω is a suitable domain, then one can define the Sobolev space $H^s(\Omega)$ as the space of Bessel potentials; roughly,

$$H^s(\Omega) = \{(1-\Delta)^{-s/2} f \mid f \in L^2(\Omega)\}.$$

Here Δ is the Laplacian and $(1 - \Delta)^{-s/2}$ is understood in terms of the spectral mapping theorem. Apart from providing a workable definition of Sobolev spaces for non-integer *s*, this definition also has particularly desirable properties under the Fourier transform that make it ideal for the study of pseudo differential operators. Using these methods on a compact Riemannian manifold, one can obtain for instance the Hodge decomposition which is the basis of Hodge theory.

Spaces of Holomorphic Functions

Hardy Spaces: The Hardy spaces are function spaces, arising in complex analysis and harmonic analysis, whose elements are certain holomorphic functions in a complex domain. Let U denote the unit disc in the complex plane. Then the Hardy space $H^2(U)$ is defined to be the space of holomorphic functions f on U such that the means

$$M_r(f) = \frac{1}{2\pi}\int_0^{2\pi} |f(re^{i\theta})|^2 \, d\theta$$

remain bounded for $r < 1$. The norm on this Hardy space is defined by

$$||f||_2 = \lim_{r\to 1} \sqrt{M_r(f)}.$$

Hardy spaces in the disc are related to Fourier series. A function f is in $H^2(U)$ if and only if

$$f(z) = \sum_{n=0}^{\infty} a_n z^n$$

where

$$\sum_{n=0}^{\infty} |a_n|^2 < \infty.$$

Thus $H^2(U)$ consists of those functions which are L^2 on the circle, and whose negative frequency Fourier coefficients vanish.

Bergman Spaces

The Bergman spaces are another family of Hilbert spaces of holomorphic functions. Let D be a bounded open set in the complex plane (or a higher dimensional complex space) and let $L^{2,h}(D)$ be the space of holomorphic functions f in D that are also in $L^2(D)$ in the sense that

$$||f||^2 = \int_D |f(z)|^2 d\mu(z) < \infty,$$

where the integral is taken with respect to the Lebesgue measurc in D. Clearly $L^{2,h}(D)$ is a subspace of $L^2(D)$; in fact, it is a closed subspace, and so a Hilbert space in its own right. This is a consequence of the estimate, valid on compact subsets K of D, that

$$\sup_{z \in K} |f(z)| \leq C_K ||f||_2,$$

which in turn follows from Cauchy's integral formula. Thus convergence of a sequence of holomorphic functions in $L^2(D)$ implies also compact convergence, and so the limit function is also holomorphic. Another consequence of this inequality is that the linear functional that evaluates a function f at a point of D is actually continuous on $L^{2,h}(D)$. The Riesz representation theorem implies that the evaluation functional can be represented as an element of $L^{2,h}(D)$. Thus, for every $z \in D$, there is a function $\eta_z \in L^{2,h}(D)$ such that

$$f(z) = \int_D f(\zeta)\overline{\eta_z(\zeta)}d\mu(\zeta)$$

for all $f \in L^{2,h}(D)$. The integrand

$$K(\zeta, z) = \overline{\eta_z(\zeta)}$$

is known as the Bergman kernel of D. This integral kernel satisfies a reproducing property

$$f(z) = \int_D f(\zeta)K(\zeta, z)d\mu(\zeta).$$

A Bergman space is an example of a reproducing kernel Hilbert space, which is a Hilbert space of functions along with a kernel $K(\zeta,z)$ that verifies a reproducing property analogous to this one. The Hardy space $H^2(D)$ also admits a reproducing kernel, known as the Szegõ kernel. Reproducing kernels are common in other areas of mathematics as well. For instance, in harmonic analysis the Poisson kernel is a reproducing kernel for the Hilbert space of square-integrable harmonic functions in the unit ball. That the latter is a Hilbert space at all is a consequence of the mean value theorem for harmonic functions.

Applications

any of the applications of Hilbert spaces exploit the fact that Hilbert spaces support generalisations of simple geometric concepts like projection and change of basis from their usual finite dimensional setting. In particular, the spectral theory of continuous self-adjoint linear operators on a Hilbert space generalises the usual spectral decomposition of a matrix, and this often plays a major role in applications of the theory to other areas of mathematics and physics.

Sturm–Liouville Theory

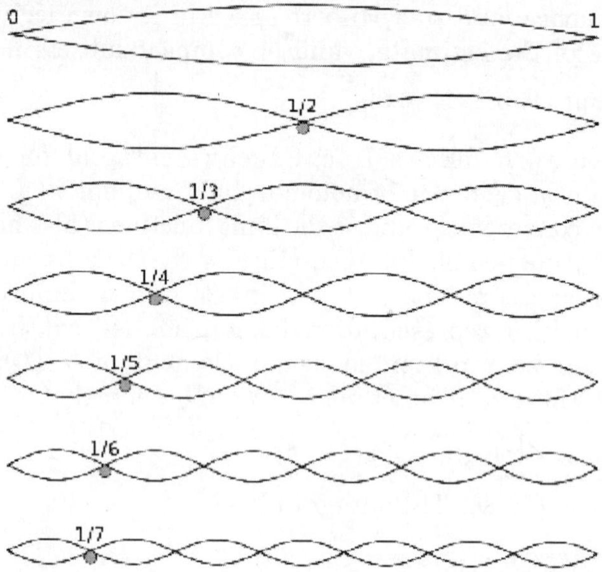

Figure 4: The overtones of a vibrating string. These are eigenfunctions of an associated Sturm–Liouville problem. The eigenvalues 1,1/2,1/3,... form the (musical) harmonic series.

In the theory of ordinary differential equations, spectral methods on a suitable Hilbert space are used to study the behaviour of eigenvalues and eigenfunctions of differential equations. For example, the Sturm–Liouville problem arises in the study of the harmonics of waves in a violin string or a drum, and is a central problem in ordinary differential equations. The problem is a differential equation of the form

$$-\frac{d}{dx}\left[p(x)\frac{dy}{dx}\right]+q(x)y=\lambda w(x)y$$

for an unknown function y on an interval $[a,b]$, satisfying general homogeneous Robin boundary conditions

$$\begin{cases}\alpha y(a)+\alpha' y'(a)=0\\ \beta y(b)+\beta' y'(b)=0.\end{cases}$$

The functions p, q, and w are given in advance, and the problem is to find the function y and constants λ for which the equation has a solution. The problem only has solutions for certain values of λ, called eigenvalues of the system, and this is a consequence of the spectral theorem for compact operators applied to the integral operator defined by the Green's function for the system. Furthermore, another consequence of this general result is that the eigenvalues λ of the system can be arranged in an increasing sequence tending to infinity.

Partial Differential Equations

Hilbert spaces form a basic tool in the study of partial differential equations. For many classes of partial differential equations, such as linear elliptic equations, it is possible to consider a generalised solution (known as a weak solution) by enlarging the class of functions. Many weak formulations involve the class of Sobolev functions, which is a Hilbert space. A suitable weak formulation reduces to a geometrical problem the analytic problem of finding a solution or, often what is more important, showing that a solution exists and is unique for given boundary data. For linear elliptic equations, one geometrical result that ensures unique solvability for a large class of problems is the Lax–Milgram theorem. This strategy forms the rudiment of the Galerkin method (a finite element method) for numerical solution of partial differential equations.

A typical example is the Poisson equation $-\Delta u = g$ with Dirichlet boundary conditions in a bounded domain Ω in R^2. The weak formulation consists of finding a function u such that, for all continuously differentiable functions v in Ω vanishing on the boundary:

$$\int_\Omega \nabla u \cdot \nabla v = \int_\Omega gv.$$

This can be recast in terms of the Hilbert space *H*1

0(Ω) consisting of functions u such that u, along with its weak partial derivatives, are square integrable on Ω, and which vanish on the boundary. The question then reduces to finding u in this space such that for all v in this space

$$a(u,v) = b(v)$$

where a is a continuous bilinear form, and b is a continuous linear functional, given respectively by

$$a(u,v) = \int_\Omega \nabla u \cdot \nabla v, \quad b(v) = \int_\Omega gv.$$

Since the Poisson equation is elliptic, it follows from Poincaré's inequality that the bilinear form a iscoercive. The Lax–Milgram theorem then ensures the existence and uniqueness of solutions of this equation. Hilbert spaces allow for many elliptic partial differential equations to be formulated in a similar way, and the Lax–Milgram theorem is then a basic tool in their analysis. With suitable modifications, similar techniques can be applied to parabolic partial differential equations and certain hyperbolic partial differential equations.

Ergodic Theory

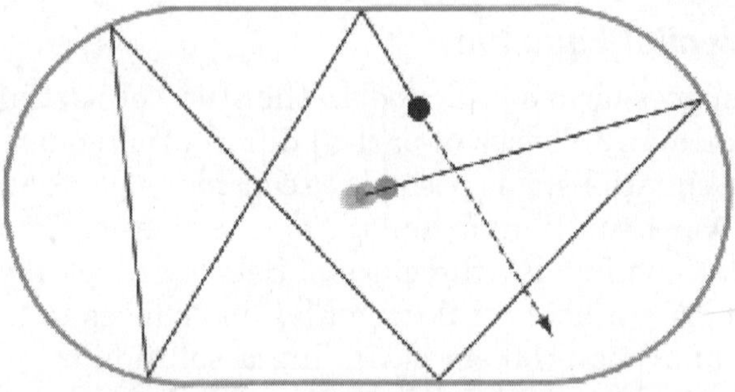

Figure 5: The path of a billiard ball in the Bunimovich stadium is described by an ergodic dynamical system.

The field of ergodic theory is the study of the long-term behaviour of chaotic dynamical systems. The protypical case of a field to which ergodic theory is applicable is that of thermodynamics in which, although the microscopic state of a system is extremely complicated—it is impossible to understand the ensemble of individual collisions between particles of matter—the average behaviour over sufficiently long time intervals is tractable.

The laws of thermodynamics are assertions about such average behaviour. In particular, one formulation of the zeroth law of thermodynamics asserts that over sufficiently long timescales, the only functionally independent measurement that one can make of a thermodynamic system in equilibrium is its total energy, in the form of temperature.

An ergodic dynamical system is one for which, apart from the energy—measured by the Hamiltonian—there are no other functionally independent conserved quantities on the phase space. More explicitly, suppose that the energy E is fixed, and let Ω_E be the subset of the phase space consisting of all states of energy E (an energy surface), and let T_t denote the evolution operator on the phase space. The dynamical system is ergodic if there are no continuous non-constant functions on Ω_E such that

$$f(T_t w) = f(w)$$

for all w on Ω_E and all time t. Liouville's theorem implies that there exists a measure μ on the energy surface that is invariant under the time translation. As a result, time translation is a unitary transformation of the Hilbert space $L^2(\Omega_E,\mu)$ consisting of square-integrable functions on the energy surface Ω_E with respect to the inner product

$$\langle f,g\rangle_{L^2(\Omega_E,\mu)} = \int_E f\bar{g}\,d\mu.$$

The von Neumann mean ergodic theorem states the following:

- If U_t is a (strongly continuous) one-parameter semigroup of unitary operators on a Hilbert space H, and P is the orthogonal projection onto the space of common fixed points of U_t, $\{x \in H \mid U_t x = x \text{ for all } t > 0\}$, then

$$Px = \lim_{T\to\infty} \frac{1}{T}\int_0^T U_t x\,dt.$$

For an ergodic system, the fixed set of the time evolution consists only of the constant functions, so the ergodic theorem implies the following: for any function $f \in L^2(\in\Omega_E,\mu)$,

$$L^2\text{–}\lim_{T\to\infty} \frac{1}{T}\int_0^T f(T_t w)\,dt = \int_{\Omega_E} f(y)\,d\mu(y).$$

That is, the long time average of an observable f is equal to its expectation value over an energy surface.

Fourier Analysis

One of the basic goals of Fourier analysis is to decompose a function into a (possibly infinite) linear combination of given basis functions: the associated Fourier series. The classical Fourier series associated to a function f defined on the interval [0,1] is a series of the form

$$\sum_{n=-\infty}^{\infty} a_n e^{2\pi i n\theta}$$

where

$$a_n = \int_0^1 f(\theta) e^{-2\pi i n\theta}\,d\theta.$$

The example of adding up the first few terms in a Fourier series for a sawtooth function is shown in the figure. The basis functions are sine waves with wavelengths λ/n(n=integer) shorter than the wavelength λ of the sawtooth itself (except for n=1, the *fundamental* wave). All basis functions have nodes at the nodes of the sawtooth, but all but the fundamental have additional nodes.

The oscillation of the summed terms about the sawtooth is called the Gibbs phenomenon.

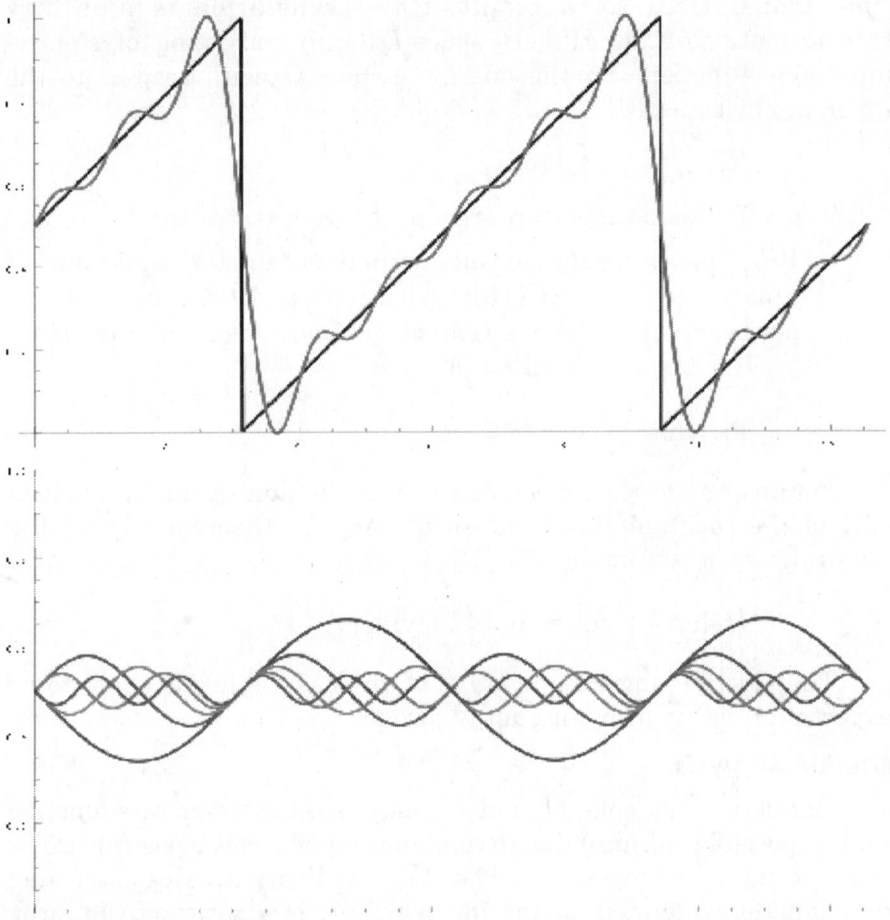

Figure 6: Superposition of sinusoidal wave basis functions (bottom) to form a sawtooth wave (top)

A significant problem in classical Fourier series asks in what sense the Fourier series converges, if at all, to the function f. Hilbert space methods provide one possible answer to this question. The functions $e_n(\theta) = e^{2\varphi in\theta}$ form an orthogonal basis of the Hilbert space $L^2([0,1])$. Consequently, any square-integrable function can be expressed as a series

$$f(\theta) = \sum_n a_n e_n(\theta), \quad a_n = \langle f, e_n \rangle$$

and, moreover, this series converges in the Hilbert space sense (that is, in the L^2 mean).

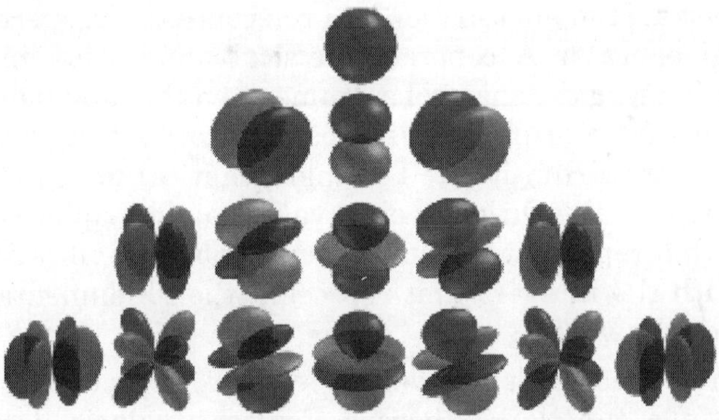

Figure 7: Spherical harmonics, an orthonourmal basis for the Hilbert space of square-integrable functions on the sphere, shown graphed along the radial direction

The problem can also be studied from the abstract point of view: every Hilbert space has an orthonourmal basis, and every element of the Hilbert space can be written in a unique way as a sum of multiples of these basis elements.

The coefficients appearing on these basis elements are sometimes known abstractly as the Fourier coefficients of the element of the space. The abstraction is especially useful when it is more natural to use different basis functions for a space such as $L^2([0,1])$. In many circumstances, it is desirable not to decompose a function into trigonometric functions, but rather into orthogonal polynomials or wavelets for instance, and in higher dimensions into spherical harmonics.

For instance, if e_n are any orthonourmal basis functions of $L^2[0,1]$, then a given function in $L^2[0,1]$ can be approximated as a finite linear combination

$$f(x) \approx f_n(x) = a_1e_1(x) + a_2e_2(x) + \cdots + a_ne_n(x)$$

The coefficients $\{a_j\}$ are selected to make the magnitude of the difference $||f - f_n||^2$ as small as possible. Geometrically, the best approximation is the orthogonal projection of f onto the subspace consisting of all linear combinations of the $\{e_j\}$, and can be calculated by

$$a_j = \int_0^1 \overline{e_j(x)} f(x)dx.$$

That this formula minimises the difference $||f - f_n||^2$ is a consequence of Bessel's inequality and Parseval's formula.

In various applications to physical problems, a function can be decomposed into physically meaningful eigenfunctions of a differential operator (typically the Laplace operator): this forms the foundation for the spectral study of functions, in reference to the spectrum of the differential operator. A concrete physical application involves the problem of hearing the shape of a drum: given the fundamental modes of vibration that a drumhead is capable of producing, can one infer the shape of the drum itself? The mathematical formulation of this question involves the Dirichlet eigenvalues of the Laplace equation in the plane, that represent the fundamental modes of vibration in direct analogy with the integers that represent the fundamental modes of vibration of the violin string.

Spectral theory also underlies certain aspects of the Fourier transform of a function. Whereas Fourier analysis decomposes a function defined on a compact set into the discrete spectrum of the Laplacian (which corresponds to the vibrations of a violin string or drum), the Fourier transform of a function is the decomposition of a function defined on all of Euclidean space into its components in the continuous spectrum of the Laplacian. The Fourier transformation is also geometrical, in a sense made precise by the Plancherel theorem, that asserts that it is an isometry of one Hilbert space (the "time domain") with another (the "frequency domain"). This isometry property of the Fourier transformation is a recurring theme in abstract harmonic analysis, as evidenced for instance by the Plancherel theorem for spherical functions occurring in noncommutative harmonic analysis.

Quantum Mechanics

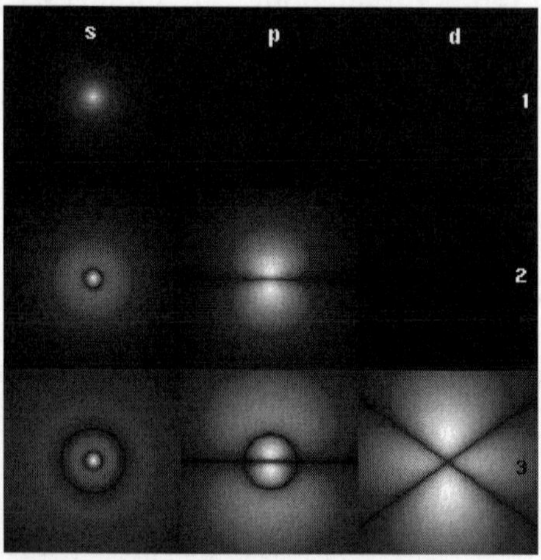

Figure 8: The orbitals of an electron in a hydrogen atom are eigenfunctions of the energy.

In the mathematically rigorous formulation of quantum mechanics, developed by Paul Dirac and John von Neumann, the possible states (more precisely, the pure states) of a quantum mechanical system are represented by unit vectors (called *state vectors*) residing in a complex separable Hilbert space, known as the state space, well defined up to a complex number of norm 1 (the phase factor). In other words, the possible states are points in the projectivisation of a Hilbert space, usually called the complex projective space.

The exact nature of this Hilbert space is dependent on the system; for example, the position and momentum states for a single non-relativistic spin zero particle is the space of all square-integrable functions, while the states for the spin of a single proton are unit elements of the two-dimensional complex Hilbert space of spinors. Each observable is represented by a self-adjointlinear operator acting on the state space. Each eigenstate of an observable corresponds to an eigenvector of the operator, and the associated eigenvalue corresponds to the value of the observable in that eigenstate. The time evolution of a quantum state is described by the Schrφdinger equation, in which the Hamiltonian, the operator corresponding to the total energy of the system, generates time evolution.

The inner product between two state vectors is a complex number known as a probability amplitude. During an ideal measurement of a quantum mechanical system, the probability that a system collapses from a given initial state to a particular eigenstate is given by the square of the absolute value of the probability amplitudes between the initial and final states. The possible results of a measurement are the eigenvalues of the operator—which explains the choice of self-adjoint operators, for all the eigenvalues must be real. The probability distribution of an observable in a given state can be found by computing the spectral decomposition of the corresponding operator.

For a general system, states are typically not pure, but instead are represented as statistical mixtures of pure states, or mixed states, given by density matrices: self-adjoint operators of trace one on a Hilbert space. Moreover, for general quantum mechanical systems, the effects of a single measurement can influence other parts of a system in a manner that is described instead by a positive operator valued measure. Thus the structure both of the states and observables in the general theory is considerably more complicated than the idealisation for pure states.

Heisenberg's uncertainty principle is represented by the statement that the operators corresponding to certain observables do not commute, and gives a specific form that the commutator must have.

Properties

Pythagorean Identity

Two vectors u and v in a Hilbert space H are orthogonal when $\langle u,v\rangle = 0$. The notation for this is u ¥" v. More generally, when S is a subset in H, the notation u ¥" S means that u is orthogonal to every element from S.

When u and v are orthogonal, one has

$$||u+v||^2 = \langle u+v,u+v\rangle = \langle u,u\rangle + 2\mathrm{Re}\langle u,v\rangle + \langle v,v\rangle = ||u||^2 + ||v||^2 .$$

By induction on n, this is extended to any family $u_1,\ldots,u_n$ of n orthogonal vectors,

$$||u_1+\cdots+u_n||^2 = ||u_1||^2 + \cdots + ||u_n||^2 .$$

Whereas the Pythagorean identity as stated is valid in any inner product space, completeness is required for the extension of the Pythagorean identity to series. A series Ó u_k of *orthogonal* vectors converges in H if and only if the series of squares of norms converges, and

$$\left|\left|\sum_{k=0}^{\infty} u_k\right|\right|^2 = \sum_{k=0}^{\infty} ||u_k||^2.$$

Furthermore, the sum of a series of orthogonal vectors is independent of the order in which it is taken.

Parallelogram Identity and Polarisation

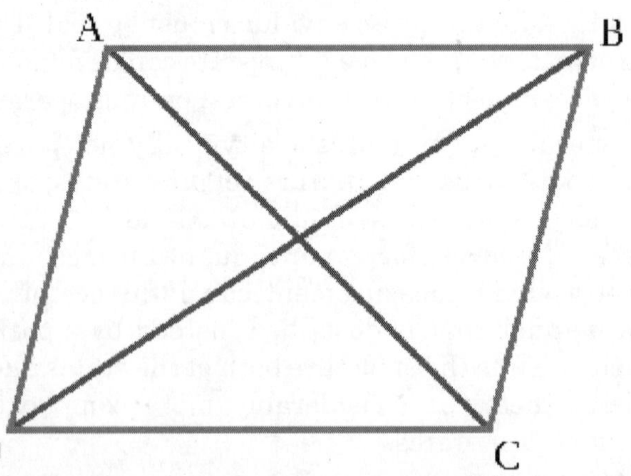

Figure 9: Geometrically, the parallelogram identity asserts that $AC^2 + BD^2 = 2(AB^2 + AD^2)$. In words, the sum of the squares of the diagonals is twice the sum of the squares of any two adjacent sides.

By definition, every Hilbert space is also a Banach space. Furthermore, in every Hilbert space the following parallelogram identity holds:

$$\|u+v\|^2 + \|u-v\|^2 = 2(\|u\|^2 + \|v\|^2).$$

Conversely, every Banach space in which the parallelogram identity holds is a Hilbert space, and the inner product is uniquely determined by the norm by the polarisation identity. For real Hilbert spaces, the polarisation identity is

$$\langle u,v\rangle = \frac{1}{4}\left(\|u+v\|^2 - \|u-v\|^2\right).$$

For complex Hilbert spaces, it is

$$\langle u,v\rangle = \frac{1}{4}\left(\|u+v\|^2 - \|u-v\|^2 + i\|u+iv\|^2 - i\|u-iv\|^2\right).$$

The parallelogram law implies that any Hilbert space is a uniformly convex Banach space.

Best Approximation

If C is a non-empty closed convex subset of a Hilbert space H and x a point in H, there exists a unique point $y \in C$ which minimises the distance between x and points in C,

$$y \in C, \quad \|x-y\| = \mathrm{dist}(x,C) = \min\{\|x-z\| : z \in C\}.$$

This is equivalent to saying that there is a point with minimal norm in the translated convex set $D = C - x$. The proof consists in showing that every minimising sequence (d_n) ," D is Cauchy (using the parallelogram identity) hence converges (using completeness) to a point in D that has minimal norm. More generally, this holds in any uniformly convex Banach space.

When this result is applied to a closed subspace F of H, it can be shown that the point $y \in F$ closest to x is characterised by

$$y \in F, \quad x - y \perp F.$$

This point y is the *orthogonal projection* of x onto F, and the mapping $P_F : x \to y$ is linear. This result is especially significant in applied mathematics, especially numerical analysis, where it forms the basis of least squares methods.

In particular, when F is not equal to H, one can find a non-zero vector v orthogonal to F (select x not in F and $v = x - y$). A very useful criterion is obtained by applying this observation to the closed subspace F generated by a subset S of H.

A subset S of H spans a dense vector subspace if (and only if) the vector 0 is the sole vector $v \in H$orthogonal to S.

Duality

The dual space H^* is the space of all continuous linear functions from the space H into the base field. It carries a natural norm, defined by

$$||\varphi||= \sup_{||x||=1, x\in H} |\varphi(x)|.$$

This norm satisfies the parallelogram law, and so the dual space is also an inner product space. The dual space is also complete, and so it is a Hilbert space in its own right.

The Riesz representation theorem affords a convenient description of the dual. To every element u of H, there is a unique element φ_u of H^*, defined by

$$\varphi_u(x) = \langle x, u\rangle.$$

The mapping $u \mapsto \varphi_u$ is an antilinear mapping from H to H^*. The Riesz representation theorem states that this mapping is an antilinear isomorphism. Thus to every element φ of the dual H^* there exists one and only one u_φ in H such that

$$\langle x, u_\varphi\rangle = \varphi(x)$$

for all $x \in H$. The inner product on the dual space H^* satisfies

$$\langle \varphi, \psi\rangle = \langle u_\psi, u_\varphi\rangle.$$

The reversal of order on the right-hand side restores linearity in φ from the antilinearity of u_φ. In the real case, the antilinear isomorphism from H to its dual is actually an isomorphism, and so real Hilbert spaces are naturally isomorphic to their own duals.

The representing vector u_φ is obtained in the following way. When $\varphi \neq 0$, the kernel $F = \ker \varphi$ is a closed vector subspace of H, not equal to H, hence there exists a non-zero vector v orthogonal to F. The vector u is a suitable scalar multiple λv of v. The requirement that $\varphi(v) = [v, u]$ yields

$$u = \langle v, v\rangle^{-1}\overline{\varphi(v)}v.$$

This correspondence $\varphi \to u$ is exploited by the bra-ket notation popular in physics. It is common in physics to assume that the inner product, denoted by $[x|y]$, is linear on the right,

$$\langle x | y\rangle = \langle y, x\rangle.$$

The result $[x \mid y]$ can be seen as the action of the linear functional $[x \mid$ (the *bra*) on the vector $\mid y]$ (the*ket*).

The Riesz representation theorem relies fundamentally not just on the presence of an inner product, but also on the completeness of the space. In fact, the theorem implies that the topological dual of any inner product space can be identified with its completion. An immediate consequence of the Riesz representation theorem is also that a Hilbert space H is reflexive, meaning that the natural map from H into its double dual space is an isomorphism.

Weakly Convergent Sequences

In a Hilbert space H, a sequence $\{x_n\}$ is weakly convergent to a vector $x \in H$ when

$$\lim_{n} \langle x_n, v \rangle = \langle x, v \rangle$$

for every $v \in H$.

For example, any orthonourmal sequence $\{f_n\}$ converges weakly to 0, as a consequence of Bessel's inequality. Every weakly convergent sequence $\{x_n\}$ is bounded, by the uniform boundedness principle.

Conversely, every bounded sequence in a Hilbert space admits weakly convergent subsequences (Alaoglu's theorem). This fact may be used to prove minimisation results for continuous convex functionals, in the same way that the Bolzano–Weierstrass theorem is used for continuous functions on R^d. Among several variants, one simple statement is as follows:

If $f : H \to R$ is a convex continuous function such that $f(x)$ tends to $+\infty$ when $\| x \|$ tends to ∞, then$f$ admits a minimum at some point $x_0 \in H$.

This fact (and its various generalisations) are fundamental for direct methods in the calculus of variations. Minimisation results for convex functionals are also a direct consequence of the slightly more abstract fact that closed bounded convex subsets in a Hilbert space H are weakly compact, since H is reflexive. The existence of weakly convergent subsequences is a special case of the Eberlein–Šmulian theorem.

Banach Space Properties

Any general property of Banach spaces continues to hold for Hilbert spaces. The open mapping theorem states that a continuous surjective linear transformation from one Banach space to another is an open mapping meaning that it sends open sets to open sets. A corollary is the bounded inverse theorem, that a continuous and bijective linear function from one Banach space to another is an isomorphism (that is, a continuous linear map whose inverse is also continuous). This theorem is considerably simpler to prove in the case of Hilbert spaces than in general Banach spaces. The open mapping theorem is equivalent to the closed graph theorem, which asserts that a function from one Banach space to another is continuous if and only if its graph is a closed set. In the case of Hilbert spaces, this is basic in the study of unbounded operators.

The (geometrical) Hahn–Banach theorem asserts that a closed convex set can be separated from any point outside it by means of a hyperplane of the Hilbert space. This is an immediate consequence of the best approximation property: if y is the element of a closed convex set F closest to x, then the separating hyperplane is the plane perpendicular to the segment xy passing through its midpoint.

Operators on Hilbert Spaces

Bounded Operators: The continuous linear operators $A : H_1 \to H_2$ from a Hilbert space H_1 to a second Hilbert space H_2 are bounded in the sense that they map bounded sets to bounded sets. Conversely, if an operator is bounded, then it is continuous. The space of such bounded linear operators has a norm, the operator norm given by

$$||A|| = \sup\{||Ax|| : ||x|| \leq 1\}.$$

The sum and the composite of two bounded linear operators is again bounded and linear. For y in H_2, the map that sends $x \in H_1$ to $[Ax, y]$ is linear and continuous, and according to the Riesz representation theorem can therefore be represented in the form

$$\langle x, A^* y\rangle = \langle Ax, y\rangle$$

for some vector A^*y in H_1. This defines another bounded linear operator $A^* : H_2 \to H_1$, the adjoint of An.. e ncan see that $A^{**} = A$.

The set B(H) of all bounded linear operators on H, together with the addition and composition operations, the norm and the adjoint operation, is a C*-algebra, which is a type of operator algebra.

An element A of B(H) is called *self-adjoint* or *Hermitian* if $A^* = A$. If A is Hermitian and $[Ax, x] \geq 0$ for every x, then A is called *non-negative*, written $A \geq 0$; if equality holds only when $x = 0$, then A is called *positive*. The set of self adjoint operators admits a partial order, in which $A \geq B$ if $A - B \geq 0$. If A has the form B^*B for some B, then A

is non-negative; if B is invertible, then A is positive. A converse is also true in the sense that, for a non-negative operator A, there exists a unique non-negative square root B such that

$$A = B^2 = B^*B.$$

In a sense made precise by the spectral theorem, self-adjoint operators can usefully be thought of as operators that are "real". An element A of B(H) is called *normal* if $A^*A = A\,A^*$. Normal operators decompose into the sum of a self-adjoint operators and an imaginary multiple of a self adjoint operator

$$A = \frac{A + A^*}{2} + i\frac{(A - A^*)}{2i}$$

that commute with each other. Normal operators can also usefully be thought of in terms of their real and imaginary parts.

An element U of B(H) is called unitary if U is invertible and its inverse is given by U^*. This can also be expressed by requiring that U be onto and $[Ux, Uy] = [x, y]$ for all x and y in H. The unitary operators form a group under composition, which is the isometry group of H.

An element of B(H) is compact if it sends bounded sets to relatively compact sets. Equivalently, a bounded operator T is compact if, for any bounded sequence $\{x_k\}$, the sequence $\{Tx_k\}$ has a convergent subsequence. Many integral operators are compact, and in fact define a special class of operators known as Hilbert–Schmidt operators that are especially important in the study of integral equations. Fredholm operators are those which differ from a compact operator by a multiple of the identity, and are equivalently characterised as operators with a finite dimensional kernel and cokernel. The index of a Fredholm operator T is defined by

$$\operatorname{index} T = \dim\ker T - \dim\operatorname{coker} T.$$

The index is homotopy invariant, and plays a deep role in differential geometry via the Atiyah–Singer index theorem.

Unbounded Operators

Unbounded operators are also tractable in Hilbert spaces, and have important applications to quantum mechanics. An unbounded operator T on a Hilbert space H is defined to be a linear operator whose domain $D(T)$ is a linear subspace of H. Often the domain $D(T)$ is a dense subspace of H, in which case T is known as a densely defined operator.

The adjoint of a densely defined unbounded operator is defined in essentially the same manner as for bounded operators. Self-adjoint unbounded operators play the role of the *observables* in the mathematical formulation of quantum mechanics. Examples of self-adjoint unbounded operators on the Hilbert space $L^2(\mathrm{R})$ are:

A suitable extension of the differential operator

$$(\mathrm{Af})(\mathrm{x}) = \mathrm{i}\frac{\mathrm{d}}{\mathrm{dx}}\mathrm{f}(\mathrm{x}),$$

where i is the imaginary unit and f is a differentiable function of compact support.

The multiplication-by-x operator:

$$(\mathrm{Bf})(\mathrm{x}) = \mathrm{xf}(\mathrm{x}).$$

These correspond to the momentum and position observables, respectively. Note that neither A nor B is defined on all of H, since in the case of A the derivative need not exist, and in the case of B the product function need not be square integrable. In both cases, the set of possible arguments form dense subspaces of $L^2(\mathrm{R})$.

Constructions

Direct Sums

Two Hilbert spaces H_1 and H_2 can be combined into another Hilbert space, called the (orthogonal) direct sum, and denoted

$$\mathrm{H}_1 \oplus \mathrm{H}_2,$$

consisting of the set of all ordered pairs (x_1, x_2) where $x_i \in H_i$, i = 1,2, and inner product defined by

$$\langle(\mathrm{x}_1,\mathrm{x}_2),(\mathrm{y}_1,\mathrm{y}_2)\rangle_{\mathrm{H}_1\oplus\mathrm{H}_2} = \langle\mathrm{x}_1,\mathrm{y}_1\rangle_{\mathrm{H}_1} + \langle\mathrm{x}_2,\mathrm{y}_2\rangle_{\mathrm{H}_2}.$$

More generally, if H_i is a family of Hilbert spaces indexed by $i \in I$, then the direct sum of the H_i, denoted

$$\bigoplus_{\mathrm{i}\in\mathrm{I}}\mathrm{H}_\mathrm{i}$$

consists of the set of all indexed families

$$\mathrm{x} = (\mathrm{x}_\mathrm{i} \in \mathrm{H}_\mathrm{i} \mid \mathrm{i} \in \mathrm{I}) \in \prod_{\mathrm{i}\in\mathrm{I}}\mathrm{H}_\mathrm{i}$$

in the Cartesian product of the H_i such that

$$\sum_{\mathrm{i}\in\mathrm{I}}\| \mathrm{x}_\mathrm{i} \|^2 < \infty.$$

The inner product is defined by

$$\langle x,y\rangle = \sum_{i\in I}\langle x_i, y_i\rangle_{H_i}.$$

Each of the H_i is included as a closed subspace in the direct sum of all of the H_i. Moreover, the H_i are pairwise orthogonal. Conversely, if there is a system of closed subspaces V_i, $i \in I$, in a Hilbert space H which are pairwise orthogonal and whose union is dense in H, then H is canonically isomorphic to the direct sum of V_i.

In this case, H is called the internal direct sum of the V_i. A direct sum (internal or external) is also equipped with a family of orthogonal projections E_i onto the ith direct summand H_i. These projections are bounded, self-adjoint, idempotent operators which satisfy the orthogonality condition

$$E_i E_j = 0, \quad i \neq j.$$

The spectral theorem for compact self-adjoint operators on a Hilbert space H states that H splits into an orthogonal direct sum of the eigenspaces of an operator, and also gives an explicit decomposition of the operator as a sum of projections onto the eigenspaces. The direct sum of Hilbert spaces also appears in quantum mechanics as the Fock space of a system containing a variable number of particles, where each Hilbert space in the direct sum corresponds to an additional degree of freedom for the quantum mechanical system. In representation theory, the Peter–Weyl theorem guarantees that any unitary representation of a compact group on a Hilbert space splits as the direct sum of finite-dimensional representations.

Tensor Products

If H_1 and H_2, then one defines an inner product on the (ordinary) tensor product as follows. On simple tensors, let

$$\langle x_1 \otimes x_2, y_1 \otimes y_2\rangle = \langle x_1, y_1\rangle\langle x_2, y_2\rangle.$$

This formula then extends by sesquilinearity to an inner product on $H_1 \otimes H_2$. The Hilbertian tensor product of H_1 and H_2, sometimes denoted by $H_1 \hat{\otimes} H_2$, is the Hilbert space obtained by completing $H_1 \otimes H_2$ for the metric associated to this inner product.

An example is provided by the Hilbert space $L^2([0, 1])$. The Hilbertian tensor product of two copies of $L^2([0, 1])$ is isometrically and linearly isomorphic to the space $L^2([0, 1]^2)$ of square-integrable functions on the square $[0, 1]^2$. This isomorphism sends a simple tensor $f_1 \otimes f_2$ to the function

$$(s,t) \mapsto f_1(s)f_2(t)$$

on the square.

This example is typical in the following sense. Associated to every simple tensor product $x_1 \otimes x_2$ is the rank one operator

$$x^* \in H_1^* \to x^*(x_1)x_2$$

from the (continuous) dual H_1^* to H_2. This mapping defined on simple tensors extends to a linear identification between $H_1 \otimes H_2$ and the space of finite rank operators from H_1^* to H_2. This extends to a linear isometry of the Hilbertian tensor product $H_1 \hat{\otimes} H_2$ with the Hilbert space $HS(H_1^*, H_2)$ of Hilbert–Schmidt operators from H_1^* to H_2.

Orthonourmal Bases

The notion of an orthonourmal basis from linear algebra generalises over to the case of Hilbert spaces. In a Hilbert space H, *an orthonourmal basis is a family* $\{e_k\}_{k \cdot B}$ *of elements of* H *satisfying the conditions:*

1. *Orthogonality*: Every two different elements of B are orthogonal: $[e_k, e_j] = 0$ for all k, j in B with$k \neq j$.
2. *Normalisation*: Every element of the family has norm $1: ||e_k|| = 1$ for all k in B.
3. *Completeness*: The linear span of the family e_k, $k \in B$, is dense in H.

A system of vectors satisfying the first two conditions basis is called an orthonourmal system or an orthonourmal set (or an orthonourmal sequence if B is countable). Such a system is always linearly independent. Completeness of an orthonourmal system of vectors of a Hilbert space can be equivalently restated as:

if $[v, e_k] = 0$ for all $k \in B$ and some $v \in H$ then $v = 0$.

This is related to the fact that the only vector orthogonal to a dense linear subspace is the zero vector, for if S is any orthonourmal set and v is orthogonal to S, then v is orthogonal to the closure of the linear span of S, which is the whole space.

Examples of orthonourmal bases include:

- the set {(1,0,0), (0,1,0), (0,0,1)} forms an orthonourmal basis of R^3 with the dot product;

- the sequence $\{f_n : n \in Z\}$ with $f_n(x) = \exp(2\pi inx)$ forms an orthonourmal basis of the complex space $L^2([0,1])$;

In the infinite-dimensional case, an orthonourmal basis will not be a basis in the sense of linear algebra; to distinguish the two, the latter basis is also called a Hamel basis. That the span of the basis vectors is dense implies that every vector in the space can be written as the sum of an infinite series, and the orthogonality implies that this decomposition is unique.

Sequence Spaces

The space ℓ^2 of square-summable sequences of complex numbers is the set of infinite sequences

$$(c_1, c_2, c_3, \ldots)$$

of complex numbers such that

$$|c_1|^2 + |c_2|^2 + |c_3|^2 + \cdots < \infty.$$

This space has an orthonourmal basis:

$$\begin{aligned} e_1 &= (1,0,0,\ldots) \\ e_2 &= (0,1,0,\ldots) \\ &\vdots \end{aligned}$$

More generally, if B is any set, then one can form a Hilbert space of sequences with index set B, defined by

$$\ell^2(B) = \left\{ x : B \xrightarrow{x} \mathbb{C} \mid \sum_{b \in B} |x(b)|^2 < \infty \right\}.$$

The summation over B is here defined by

$$\sum_{b \in B} |x(b)|^2 = \sup \sum_{n=1}^{N} |x(b_n)|^2$$

the supremum being taken over all finite subsets of B. It follows that, in order for this sum to be finite, every element of $\ell^2(B)$ has only countably many nonzero terms. This space becomes a Hilbert space with the inner product

$$\langle x, y \rangle = \sum_{b \in B} x(b)\overline{y(b)}$$

for all x and y in $\ell^2(B)$. Here the sum also has only countably many nonzero terms, and is unconditionally convergent by the Cauchy–Schwarz inequality. An orthonourmal basis of $\ell^2(B)$ is indexed by the set B, given by

$$e_b(b') = \begin{cases} 1 & \text{if } b = b' \\ 0 & \text{otherwise.} \end{cases}$$

Bessel's Inequality and Parseval's Formula

Let $f_1, ..., f_n$ be a finite orthonourmal system in H. For an arbitrary vector x in H, let

$$y = \sum_{j=1}^{n} \langle x, f_j \rangle f_j.$$

Then $[x, f_k] = [y, f_k]$ for every $k = 1, ..., n$. It follows that $x - y$ is orthogonal to each f_k, hence$x - y$ is orthogonal to y. Using the Pythagorean identity twice, it follows that

$$||x||^2 = ||x - y||^2 + ||y||^2 \geq ||y||^2 = \sum_{j=1}^{n} |\langle x, f_j \rangle|^2 .$$

Let $\{f_i\}$, $i \in I$, be an arbitrary orthonourmal system in H. Applying the preceding inequality to every finite subset J of I gives the *Bessel inequality*

$$\sum_{i \in I} |\langle x, f_i \rangle|^2 \leq ||x||^2, \quad x \in H$$

(according to the definition of the sum of an arbitrary family of non-negative real numbers).

Geometrically, Bessel's inequality implies that the orthogonal projection of x onto the linear subspace spanned by the f_i has norm that does not exceed that of x. In two dimensions, this is the assertion that the length of the leg of a right triangle may not exceed the length of the hypotenuse.

Bessel's inequality is a stepping stone to the more powerful Parseval identity which governs the case when Bessel's inequality is actually an equality. If $\{e_k\}_{k \in B}$ is an orthonourmal basis of H, then every element x of H may be written as

$$x = \sum_{k \in B} \langle x, e_k \rangle e_k.$$

Even if B is uncountable, Bessel's inequality guarantees that the expression is well-defined and consists only of countably many nonzero terms. This sum is called the *Fourier expansion* of x, and the individual coefficients $[x,e_k]$ are the *Fourier coefficients* of x. Parseval's formula is then

$$||x||^2 = \sum_{k \in B} |\langle x, e_k \rangle|^2 .$$

Conversely, if $\{e_k\}$ is an orthonourmal set such that Parseval's identity holds for every x, then $\{e_k\}$ is an orthonourmal basis.

Hilbert Dimension

As a consequence of Zorn's lemma, *every* Hilbert space admits an orthonourmal basis; furthermore, any two orthonourmal bases of the same space have the same cardinality, called the Hilbert dimension of the space. For instance, since $\ell^2(B)$ has an orthonourmal basis indexed by B, its Hilbert dimension is the cardinality of B (which may be a finite integer, or a countable or uncountable cardinal number).

As a consequence of Parseval's identity, if $\{e_k\}_{k \cdot B}$ is an orthonourmal basis of H, then the map $\Phi : H \to \ell^2(B)$ defined by $\Phi(x) = ([x,e_k])_{k \in B}$ is an isometric isomorphism of Hilbert spaces: it is a bijective linear mapping such that

$$\langle \Phi(x), \Phi(y) \rangle_{\ell^2(B)} = \langle x, y \rangle_H$$

for all x and y in H. The cardinal number of B is the Hilbert dimension of H. Thus every Hilbert space is isometrically isomorphic to a sequence space $\ell^2(B)$ for some set B.

Separable Spaces

A Hilbert space is separable if and only if it admits a countable orthonourmal basis. All infinite-dimensional separable Hilbert spaces are therefore isometrically isomorphic to ℓ^2.

In the past, Hilbert spaces were often required to be separable as part of the definition. Most spaces used in physics are separable, and since these are all isomorphic to each other, one often refers to any infinite-dimensional separable Hilbert space as "*the* Hilbert space" or just "Hilbert space". Even in quantum field theory, most of the Hilbert spaces are in fact separable, as stipulated by the Wightman axioms.

However, it is sometimes argued that non-separable Hilbert spaces are also important in quantum field theory, roughly because the systems in the theory possess an infinite number of degrees of freedom and any infinite Hilbert tensor product (of spaces of dimension greater than one) is non-separable. For instance, a bosonic field can be naturally thought of as an element of a tensor product whose factors represent harmonic oscillators at each point of space. From this perspective, the natural state space of a boson might seem to be a non-separable space. However, it is only a small separable subspace of the full tensor product that can contain physically meaningful fields (on which the observables can be defined). Another non-separable Hilbert space models the state of an infinite collection of particles in an unbounded region of space. An orthonourmal basis of the space is indexed by the density of the particles, a continuous parameter, and since the set of possible densities is uncountable, the basis is not countable.

Orthogonal Complements and Projections

If S *is a subset of a Hilbert space* H, *the set of vectors orthogonal to* S *is defined by*

$$S^{\perp} = \{x \in H : \langle x,s \rangle = 0 \ \forall s \in S\}.$$

$S^{\perp}$ is a closed subspace of H (can be proved easily using the linearity and continuity of the inner product) and so forms itself a Hilbert space. If V is a closed subspace of H, then $V^{\perp}$ is called the *orthogonal complement* of V. In fact, every x in H can then be written uniquely as $x = v + w$, with v in V and w in $V^{\perp}$. Therefore, H is the internal Hilbert direct sum of V and $V^{\perp}$.

The linear operator $P_V : H \to H$ which maps x to v is called the *orthogonal projection* onto V. There is a onaturalne-to-one correspondence between the set of all closed subspaces of H and the set of all bounded self-adjoint operators P such that $P^2 = P$. Specifically,

Theorem. The orthogonal projection P_V is a self-adjoint linear operator on H of norm ≤ 1 with the property $P^2{}_V = P_V$. Moreover, any self-adjoint linear operator E such that $E^2 = E$ is of the form P_V, where V is the range of E. For every x in H, $P_V(x)$ is the unique element v of V which minimises the distance $||x - v||$.

Projections P_U and P_V are called mutually orthogonal if $P_U P_V = 0$. This is equivalent to U and V being orthogonal as subspaces of H. The sum of the two projections P_U and P_V is a projection only if U and V are orthogonal to each other, and in that case $P_U + P_V = P_{U+V}$. The composite PUPV is generally not a projection; in fact, the composite is a projection if and only if the two projections commute, and in that case $P_U P_V = P_U n_V$.

By restricting the codomain to the Hilbert space V, the orthogonal projection PV gives rise to a projection mapping $\pi : H \to V$; it is the adjoint of the inclusion mapping

$$i_V : V \to H,$$

meaning that

$$\langle i_V x, y \rangle = \langle x, P_V y \rangle$$

for all $x \in H$ and $y \in V$.

The operator norm of a projection P onto a non-zero closed subspace is equal to one:

$$||P|| = \sup_{x \in H, x \neq 0} \frac{||Px||}{||x||} = 1.$$

Every closed subspace V of a Hilbert space is therefore the image of an operator P of norm one such that $P^2 = P$. In fact this property characterises Hilbert spaces:

A Banach space of dimension higher than 2 is (isometrically) a Hilbert space if and only if, to every closed subspace V, there is an operator P_V of norm one whose image is V such that $P_V^2 = P_V$.

While this result characterises the metric structure of a Hilbert space, the structure of a Hilbert space as a topological vector space can itself be characterised in terms of the presence of complementary subspaces:

- A Banach space X is topologically and linearly isomorphic to a Hilbert space if and only if, to every closed subspace V, there is a closed subspace W such that X is equal to the internal direct sum $V \oplus W$.

The orthogonal complement satisfies some more elementary results. It is a monotone function in the sense that if $U \subset V$, then $V^{\perp} \subseteq U^{\perp}$ with equality holding if and only if V is contained in the closure of U. This result is a special case of the Hahn–Banach theorem. The closure of a subspace can be completely characterised in terms of the orthogonal complement: If V is a subspace of H, then the closure of V is equal to $V^{\perp\perp}$. The orthogonal complement is thus a Galois connection on the partial order of subspaces of a Hilbert space. In general, the orthogonal complement of a sum of subspaces is the intersection of the orthogonal complements: $\left(\sum_i V_i\right)^{\perp} = \bigcap_i V_i^{\perp}$.

If the V_i are in addition closed, then $\overline{\sum_i V_i^{\perp}} = \left(\bigcap_i V_i\right)^{\perp}$.

Spectral Theory

There is a well-developed spectral theory for self-adjoint operators in a Hilbert space, that is roughly analogous to the study of symmetric matrices over the reals or self-adjoint matrices over the complex numbers. In the same sense, one can obtain a "diagonalisation" of a self-adjoint operator as a suitable sum (actually an integral) of orthogonal projection operators.

The spectrum of an operator T, denoted $\sigma(T)$ is the set of complex numbers λ such that $T - \lambda$ lacks a continuous inverse. If T is bounded, then the spectrum is always a compact set in the complex plane, and lies inside the disc $|z| \leq ||T||$. If T is self-adjoint, then the spectrum is real. In fact, it is contained in the interval $[m,M]$ where

$$m = \inf_{||x||=1} \langle Tx, x \rangle, \quad M = \sup_{||x||=1} \langle Tx, x \rangle.$$

Moreover, m and M are both actually contained within the spectrum.

The eigenspaces of an operator T are given by

$$H_\lambda = \ker(T - \lambda).$$

Unlike with finite matrices, not every element of the spectrum of T must be an eigenvalue: the linear operator $T - \lambda$ may only lack an inverse because it is not surjective. Elements of the spectrum of an operator in the general sense are known as *spectral values*. Since spectral values need not be eigenvalues, the spectral decomposition is often more subtle than in finite dimensions.

However, the spectral theorem of a self-adjoint operator T takes a particularly simple form if, in addition, T is assumed to be a compact operator. The spectral theorem for compact self-adjoint operators states:

- A compact self-adjoint operator T has only countably (or finitely) many spectral values. The spectrum of T has no limit point in the complex plane except possibly zero. The eigenspaces of T decompose H into an orthogonal direct sum:

$$H = \bigoplus_{\lambda \in \sigma(T)} H_\lambda.$$

Moreover, if E_λ denotes the orthogonal projection onto the eigenspace H_λ, then

$$T = \sum_{\lambda \in \sigma(T)} \lambda E_\lambda$$

where the sum converges with respect to the norm on $B(H)$.

This theorem plays a fundamental role in the theory of integral equations, as many integral operators are compact, in particular those that arise from Hilbert–Schmidt operators.

The general spectral theorem for self-adjoint operators involves a kind of operator-valued Riemann–Stieltjes integral, rather than an infinite summation. The *spectral family* associated to T associates to each real number λ an operator E_λ, which is the projection onto the nullspace of the operator$(T-\lambda)^+$, where the positive part of a self-adjoint operator is defined by

$$A^+ = \frac{1}{2}\left(\sqrt{A^2} + A\right).$$

The operators E_λ are monotone increasing relative to the partial order defined on self-adjoint operators; the eigenvalues correspond precisely to the jump discontinuities. One has the spectral theorem, which asserts

$$T = \int_{\mathbb{R}} \lambda dE_\lambda.$$

The integral is understood as a Riemann–Stieltjes integral, convergent with respect to the norm on B(*H*). In particular, one has the ordinary scalar-valued integral representation

$$\langle Tx, y\rangle = \int_{\mathbb{R}} \lambda d\langle E_\lambda x, y\rangle.$$

A somewhat similar spectral decomposition holds for normal operators, although because the spectrum may now contain non-real complex numbers, the operator-valued Stieltjes measure dE_λ must instead be replaced by a resolution of the identity.

A major application of spectral methods is the spectral mapping theorem, which allows one to apply to a self-adjoint operator T any continuous complex function f defined on the spectrum of T by forming the integral

$$f(T) = \int_{\sigma(T)} f(\lambda) dE_\lambda.$$

The resulting continuous functional calculus has applications in particular to pseudo differential operators.

The spectral theory of *unbounded* self-adjoint operators is only marginally more difficult than for bounded operators. The spectrum of an unbounded operator is defined in precisely the same way as for bounded operators: λ is a spectral value if the resolvent operator

$$R_\lambda = (T - \lambda)^{-1}$$

fails to be a well-defined continuous operator. The self-adjointness of T still guarantees that the spectrum is real. Thus the essential idea of working with unbounded operators is to look instead at the resolvent R_λ where λ is non-real. This is a *bounded* normal operator, which admits a spectral representation that can then be transferred to a spectral representation of T itself. A similar strategy is used, for instance, to study the spectrum of the Laplace operator: rather than address the operator directly, one instead looks as an associated resolvent such as a Riesz potential or Bessel potential.

A precise version of the spectral theorem which holds in this case is:

Given a densely defined self-adjoint operator T on a Hilbert space H, there corresponds a uniqueresolution of the identity E on the Borel sets of R, such that

$$\langle Tx, y\rangle = \int_{\mathbb{R}} \lambda dE_{x,y}(\lambda)$$

for all $x \in D(T)$ and $y \in H$. The spectral measure E is concentrated on the spectrum of T.

There is also a version of the spectral theorem that applies to unbounded normal operators.

Banach Algebra

In mathematics, especially functional analysis, a Banach algebra, named after Stefan Banach, is an associative algebra A over the real or complex numbers which at the same time is also a Banach space. The algebra multiplication and the Banach space norm are required to be related by the following inequality:

$$\forall x, y \in A : ||xy|| \leq ||x||\,||y||$$

(i.e., the norm of the product is less than or equal to the product of the norms). This ensures that the multiplication operation is continuous. This property is found in the real and complex numbers; for instance, $|-6\times 5| \leq |-6|\times|5|$.

If in the above we relax Banach space to normed space the analogous structure is called a normed algebra.

A Banach algebra is called "unital" if it has an identity element for the multiplication whose norm is 1, and "commutative" if its multiplication is commutative. Any Banach algebra A (whether it has anidentity element or not) can be embedded isometrically into a unital Banach algebra A_e so as to form a closed ideal of A_e. Often one assumes *a priori* that the algebra under consideration is unital: for one can develop much of the theory by considering A_e and then applying the outcome in the original algebra. However, this is not the case all the time. For example, one cannot define all the trigonometric functions in a Banach algebra without identity.

The theory of real Banach algebras can be very different from the theory of complex Banach algebras. For example, the spectrum of an element of a complex Banach algebra can never be empty, whereas in a real Banach algebra it could be empty for some elements.

Banach algebras can also be defined over fields of p-adic numbers. This is part of p-adic analysis.

Spectral Theory

Unital Banach algebras over the complex field provide a general setting to study spectral theory. The spectrum of an element $x \in A$, denoted by $\sigma(x)$, consists of all those complex scalars λ such that $x-\lambda 1$ is not invertible in A. The spectrum of any element x is a closed subset of the closed disc in C with radius $||x||$ and centre 0, and thus is compact. Moreover, the spectrum $\sigma(x)$ of an element x is non-empty and satisfies the spectral radius formula:

$$\sup\{|\lambda| : \lambda \in \sigma(x)\} = \lim_{n\to\infty} ||x^n||^{1/n}.$$

Given $x \in A$, the holomorphic functional calculus allows to define $f(x) \in A$ for any function f holomorphic in a neighbourhood of $\sigma(x)$. Furthermore, the spectral mapping theorem holds:

$$\sigma\big(f(x)\big) = f\big(\sigma(x)\big).$$

When the Banach algebra A is the algebra L(X) of bounded linear operators on a complex Banach space X (e.g., the algebra of square matrices), the notion of the spectrum in A coincides with the usual one in the operator theory. For $f \in C(X)$ (with a compact Hausdorff space X), one sees that:

$$\sigma(f) = \{f(t) : t \in X\}.$$

The norm of a normal element x of a C*-algebra coincides with its spectral radius. This generalises an analogous fact for normal operators.

Let A be a complex unital Banach algebra in which every non-zero element x is invertible (a division algebra). For every $a \in A$, there is $\lambda \in$ C such that $a \in \lambda 1$ is not invertible (because the spectrum of a is not empty) hence $a = \lambda 1$: this algebra A is naturally isomorphic to C (the complex case of the Gelfand-Mazur theorem).

Ideals and Characters

Let A be a unital *commutative* Banach algebra over C. Since A is then a commutative ring with unit, every non-invertible element of A belongs to some maximal ideal of A. Since a maximal ideal $\mathfrak{m}$ in A is closed, $A/\mathfrak{m}$ is a Banach algebra that is a field, and it follows from the Gelfand-Mazur theorem that there is a bijection between the set of all maximal ideals of A and the set $\Delta(A)$ of all nonzero homomorphisms from A to C. The set $\Delta(A)$ is called the "structure space" or "character space" of A, and its members "characters."

A character χ is a linear functional on A which is at the same time multiplicative, $\chi(ab) = \chi(a)\,\chi(b)$, and satisfies $\chi(1) = 1$. Every character is automatically continuous from A to C, since the kernel of a character is a maximal ideal, which is closed. Moreover, the norm (*i.e.*, operator norm) of a character is one. Equipped with the topology of pointwise convergence on A (*i.e.*, the topology induced by the weak-* topology of A^*), the character space, $\Delta(A)$, is a Hausdorff compact space.

For any $x \in A$,

$$\sigma(\mathrm{x}) = \sigma(\hat{\mathrm{x}})$$

where $\hat{\mathrm{x}}$ is the Gelfand representation of x defined as follows: $\hat{\mathrm{x}}$ is the continuous function from $\Delta(A)$ to C given by $\hat{\mathrm{x}}(\chi) = \chi(\mathrm{x})$..

The spectrum of $\hat{\mathrm{x}}$ in the formula above, is the spectrum as element of the algebra $C(\Delta(A))$ of complex continuous functions on the compact space $\Delta(A)$. Explicitly,

$$\sigma(\hat{\mathrm{x}}) = \{\chi(\mathrm{x}) : \chi \in \Delta(\mathrm{A})\}$$

As an algebra, a unital commutative Banach algebra is semisimple (i.e., its Jacobson radical is zero) if and only if its Gelfand representation has trivial kernel. An important example of such an algebra is a commutative C*-algebra. In fact, when A is a commutative unital C*-algebra, the Gelfand representation is then an isometric *-isomorphism between A and $C(\Delta(A))$.

Bibliography

Aleksandrov, P.: *Combinatorial Topology*, Graylock Press, Baltimore, MD, 1956-60.

Alexandroff, Paul: *Elementary Concepts of Topology,* Dover, Mineola, NY, 1961.

Anton, Howard: *Elementary Linear Algebra*, John Wiley & Sons, NJ, 1987.

Armstrong, M.A.: *Basic Topology,* Springer-Verlag, New York, 1983.

Arrow, J.: *Social Choice and Individual Values*, Wiley, NJ, 1963.

Artin, Emil: *Introduction to Algebraic Topology,* Charles E. Merrill, Columbus, OH, 1969.

Auslander, Louis and Mackenzie, Robert E.: *Introduction to Differentiable,* Dover, NY, 1977.

Bachman, George; Narici, Lawrence; Beckenstein, Edward: *Fourier and Wavelet Analysis*, Springer-Verlag, New York, 2000.

Ball, W.W.: *Mathematical Recreations and Essays*, MacMillan, NJ, 1962.

Bing, R.H.: *Elementary Point Set Topology,* Mathematical Association of America, Washington, DC, 1960.

Blackett, Donald W.: *Elementary Topology: A Combinatorial and Algebraic Approach*, Academic Press, New York, 1982.

Bourbaki, Nicolas: *Elements of Mathematics: General Topology,* Springer-Verlag, New York, 1989.

Bourbaki, Nicolas: *Spectral Theories, Elements of mathematics*, Springer-Verlag, Berlin, 1986.

Bridgman, P. W.: *Dimensional Analysis*, Yale University Press, London, 1931.

Brocker, Th. and Janich, Klaus: *Introduction to Differential Topology,* Cambridge University Press, New York, 1982.

Carpenter, T.P., Franke, M.L. & Levi, L.: *Thinking Mathematically: Integrating Arithmetic and Algebra in Elementary School,* Heinemann, Portsmouth, NH, 2003.

Casey, John: *The Elements of Euclid,* Figgis, and Co., Hodges, 1890.

Chillingworth, D.R.J.: *Differential Topology with a View to Applications,* Pitman, Brooklyn, NY, 1976.

Chinn, William G. and Steenrod, Norman E.: *First Concepts of Topology,* Mathematical Association of America, Washington, DC, 1966.

Corry, Leo: *Modern Algebra and the Rise of Mathematical Structures,* Birkhäuser, Switzerland, and Boston, 1996.

Courant, Richard; Robbins, Herbert: *What is Mathematics?*, Oxford University Press, Oxford, 1978.

Crowe, Michael J.: *A History of Vector Analysis: The Evolution of the Idea of a Vertical System,* Notre Dame University Press, London, 1967.

Crowell, Richard H. and Fox, Ralph H.: *Introduction to Knot Theory,* Springer-Verlag, New York, 1977.

Dieudonne, Jean: *A History of Algebraic and Differential Topology 1900–1960,* Birkhauser, Switzerland, 1989.

Dirac, P.A.M.: *The Principles of Quantum Mechanics*, Clarendon Press, Oxford, 1930.

Dugundji, James: *Topology,* Allyn and Bacon, Boston, MA, 1966.

Duren, P.: *Theory of Hp-Spaces*, Academic Press, New York, 1970.

Firby, P.A. and Gardiner, C.F.: *Surface Topology,* Ellis Horwood, New York, 1982.

Fuks, D.B. and Rokhlin, V.A.: *Beginner's Course in Topology: Geometric Chapters,* Springer-Verlag, New York, 1984.

Gamelin, Theodore W. and Greene, Robert E.: *Introduction to Topology,* Saunders College, Philadelphia, PA, 1983.

Gemignani, Michael C.: *Elementary Topology*, Dover, Mineola, NY, 1990.

Giblin, P.J. Graphs: *Surfaces and Homology: An Introduction to Algebraic Topology*, Chapman and Hall, New York, 1981.

Goult, R.J.; Hoskins, R.F.; Milner, J.A.; Pratt, M.J.: *Computational Methods in Linear Algebra*, Wiley, NJ, 1975.

Gramain, Andre: *Topology of Surfaces,* BCS Associates, Moscow, ID, 1984.

Grattan-Guinness, I.: *Companion Encyclopedia of the History and Philosophy of the Mathematical Sciences,* Routledge, London, 1994.

Greenberg, Marvin Jay and Harper, John R.: *Lectures on Algebraic Topology*, W.A. Benjamin, Reading, MA, 1981.

Greenes, C.E. & Rubenstein, R.: *Algebra and Algebraic Thinking in School Mathematics: 70th Yearbook,* National Council of Teachers of Mathematics, Reston, VA, 2008.

Guillemin, Victor and Pollack, Alan: *Differential Topology,* Prentice Hall, Englewood Cliffs, NJ, 1974.

Hausdorff, Felix: *Set Theory*, Chelsea, New York, 1957.

Hawkins, Thomas W.: *Emergence of Lie's Theory of Groups: An Essay in the History of Mathematics, 1869–1926,* Springer, Berlin, 2002.

Henle, Michael: *A Combinatorial Introduction to Topology,* W.H. Freeman, New York, 1979.

Hirsch, Morris W.: *Differential Topology,* Springer-Verlag, New York, 1976.

Høyrup, Jens: *Lengths, Widths, Surfaces: A Portrait of Old Babylonian Algebra and Its Kin,* Springer, Berlin and New York, 2002.

Hurewicz, Witold and Wallman, Henry: *Dimension Theory,* Princeton University Press, Princeton, NJ, 1941.

Janich, Klaus: *Topology,* Springer-Verlag, New York, 1984.

Kaplansky, Irving: *Set Theory and Metric Spaces*, Chelsea, New York, 1977.

Kaput, J.J., Carraher, D.W. & Blanton. M.L.: *Algebra in the Early Grades*, Lawrence Erlbaum Associates, New York, 2008.

Kelley, John L.: *General Topology*, Springer-Verlag, New York, 1975.

Klein, J.: *Greek Mathematical Thought and the Origins of Algebra,* MIT Press, Cambridge, Mass, 1968.

Kosniowski, Czes: *A First Course in Algebraic Topology,* Cambridge University Press, New York, 1980.

Kuratowski, K.: *Topology*, Academic Press, New York, 1966.

Lam, Lay Yong, and T. S. Ang.: *Fleeting Footsteps: Tracing the Conception of Arithmetic and Algebra in Ancient China,* World Scientific Publishing, Singapore, 1992.

Lefschetz, Solomon: *Topology*, Chelsea, New York, 1956.

Massey, William S.: *Algebraic Topology: An Introduction,* Springer-Verlag, New York, 1977.

Mehrtens, Herbert: *Die Entstehung der Verbandstheorie,* Germany: Gerstenberg, Hildesheim, 1979.

Milnor, John W.: *Topology from the Differentiable Viewpoint,* University Press of Virginia, Charlottesville, VA, 1965.

Moses, R.P. & Cobb, C.E.: *Radical Equations: Civil Rights from Mississippi to the Algebra Project*, Beacon Press, Boston, MA, 2002.

Munkres, James R.: *Topology: A First Course,* Prentice Hall, Englewood Cliffs, NJ, 1975.

Naber, Gregory L.: *Topological Methods in Euclidean Spaces,* Cambridge University Press, New York, 1980.

Nesselmann, G. H. F.: *Versuch einer kritischen Geschichte der Algebra,* Minerva, Germany, 1969.

Newman, M.H.A: *Elements of the Topology of Plane Sets of Points,* Cambridge University Press, New York, NY, 1964.

Novy, Lubos: *The Origins of Modern Algebra,* Academia, Prague, 1973.

Oxtoby, John C.: *Measure and Category: A Survey of the Analogies between Topological and Measure Spaces*, Springer-Verlag, New York, NY, 1980.

Polya, G.: *Mathematics and Plausible Reasoning: Volume II Patterns of Plausible Inference,* Princeton University Press, NJ, 1954.

Pontrjagin, Lev S.: *Topological Groups*, Gordon and Breach, New York, 1966.

Pycior, Helena M.: *Symbols, Impossible Numbers, and Geometric Entanglements: British Algebra through the Commentaries on Newton's Universal Arithmetick,* Cambridge University Press, New York, 1997.

Scholz, Erhard: *Geschichte der Algebra: Eine Ausführung,* Wissenschaftsverlag, Mannheim, 1990.

Schweitzer, K.: *Teachers Engaged in Research: Inquiry into Mathematics Classrooms, Grades Pre-K-2,* Information Age Publishing, Greenwich, CT, 2006.

Seifert, Herbert and Threlfall, W. A.: *Textbook of Topology,* Academic Press, New York, 1980.

Sierpinski, W.: *General Topology,* University of Toronto Press, Toronto, 1952.

Simmons, George F.: *Introduction to Topology and Modern Analysis,* Robert E. Krieger, Melbourne, FL, 1983.

Spanier, Edwin H.: *Algebraic Topology,* McGraw-Hill, New York, 1966.

Stedall, Jacqueline A.: *A Discourse Concerning Algebra: English Algebra to 1685,* Oxford University Press, Oxford, 2002.

Steen, Lynn Arthur and Seebach, J. Arthur, Jr.: *Counterexamples in Topology*, Springer-Verlag, New York, 1978.

Stillwell, John: *Classical Topology and Combinatorial Group Theory,* Springer-Verlag, New York, 1980.

Vick, James: *Homology Theory: An Introduction to Algebraic Topology,* Academic Press, New York, 1973.

Wilkinson, J. H.: *The Algebraic Eigenvalue Problem*, Oxford University Press, Oxford, 1998.

Willard, Stephen: *General Topology,* Addison-Wesley, Reading, MA, 1970.

Wussing, Hans: *The Genesis of the Abstract Group Concept: A Contribution to the History of the Origin of Abstract Group Theory,* MIT Press, Cambridge, Mass, 1984.

Index